23rd InterAmerican Federation of ChEmical Engineering Congress

Topical Conference at the 2008 AIChE Annual Meeting

Philadelphia, Pennsylvania
16 - 21 November 2008

ISBN: 978-1-61567-215-8

Printed from e-media with permission by:

Curran Associates, Inc.
57 Morehouse Lane
Red Hook, NY 12571
www.proceedings.com

Some format issues inherent in the e-media version may also appear in this print version.

Printed by Curran Associates, Inc. (2009)

AIChE
3 Park Avenue
New York, NY 10016-5991

Phone: (203) 702-7660
Fax: (203) 775-5177

www.aiche.org

TABLE OF CONTENTS

256a Mixing in Chemical Engineering

Carlos Cabrera, UOP LLC, 25 East Algonquin Road, P.O. Box 5017, Des Plaines, IL 60017-5017

Mixing plays a critical role in nearly every chemical process, and continues to provide challenging problems for chemical engineers to solve. Dr. James Oldshue made his indelible mark on society through his work in these technologies. His ability to develop solutions helped set the industry definition of mixing innovation. We'll look at some recent mixing problems faced by UOP to illustrate the breadth of issues encountered in industry today and introduce some of the novel approaches our engineers are using.

As we enter AIChE's second century, the workplace is becoming more globalized and the demographics of our profession are changing rapidly. These trends create cultural mixing challenges that can test even the most innovative managers and leaders. UOP has a long history of collaborating with international customers and partners, as well as bringing together scientists and engineers from around the world to work together to develop new technologies. We'll look at some examples of business practices that promote this cultural and societal mixing.

Design of a Hybrid Distillation-Pervaporation Bio-Ethanol Purification Process Using Conceptual Design and Rigorous Simulation Tools

Patricia M. Hoch,[a] José Espinosa,[b]

[a]PLAPIQUI-UNS-CONICET, Camino La Carrindanga km 7-8000, Bahía Blanca, Argentina
[b]INGAR-CONICET-UNL, Avellaneda 3657, S3002 GJC Santa Fe, Argentina

Prepared for presentation at AIChE Annual Meeting 2008
Session: #316 - Renewable and Sustainable Fuel, Products, and Processes: Practice, Science and Advances in North, Central, and South America (TJ001)

Abstract

Optimization of both design and operation variables of a bio-ethanol purification plant intended to produce fuel grade ethanol is a challenge to make the bio-fuel a realistic alternative in the energy market. In this work, we propose an evolutionary optimization procedure which intensively uses both conceptual and rigorous models for the design and simulation of unit operations involved in a bio-ethanol purification process. While conceptual models are used to determine initial values for design and operating variables of a given operation unit, rigorous simulation departing from initial estimates generated at conceptual design level enables to remove simplifying assumptions, interconnect equipments and calculate operating and investment costs.

Once an initial design of the purification plant is obtained, opportunities of improvement are easily recognized and then tested by performing the design and simulation steps until a cost-effective bio-ethanol purification plant is achieved.

Finding initial values for the column design is made easier within the conceptual design environment, which is used to obtain an initial estimation of the total number of trays of the main distillation column, placement of feed and side streams, and steam flow rate in the simulation environment. In addition, convergence of the rigorous model is enhanced by using initial estimates of internal profiles generated at the conceptual design level despite of the highly non-ideal behavior of the multicomponent azeotropic mixture.

The advantage of the method is that the systematic use of conceptual models allows the designer to capture the main characteristics of each operation involved in the process.

The main results obtained show a minimum in the operation costs corresponding to a minimum in the steam flow rate of the hybrid column. This minimum in steam flow rate can be only explained by the presence of the fusel component, which influences both the energy demand and feasible products of the hybrid column. Therefore, designs based on the binary system ethanol-water do not represent the system behavior in an accurate way. From this consideration emerges the importance of properly determining the amount of trace components that enter the purification process.

From the analysis of the results obtained it is also clear that the investment cost corresponding to the membrane sector is high enough to promote future research efforts in testing both selectivity and flux behavior of other commercial pervaporation membranes.

Keywords: *Bio-ethanol Purification, Evolutionary Optimization, Conceptual Design, Rigorous Simulation.*

1. Introduction

There is great incentive in developing bio fuel processes, and optimization constitutes a tool that allows the improvement of process conditions and structure. Bio-ethanol produced from corn is a bio-fuel of interest, but the complexity of the production plant makes it difficult to propose a general superstructure taking into account all possibilities, and the high non-linearity of the corresponding models requires rigorous solution methods. In this work, we propose an evolutionary optimization procedure which uses intensively both conceptual and rigorous models for the design and simulation of unit operations.

The bio-ethanol purification plant comprises a separation sector with two distillation columns where both the reboilers are replaced by live steam feeds, a decanter and a pervaporation unit, used to obtain fuel grade ethanol. The plant is integrated with two other sectors, a fusel plant with several liquid-liquid separators, and an evaporation plant, where a triple effect evaporator, a centrifuge and a rotary dryer are used to obtain an animal feed co-product rich in proteins. Given the plant complexity and the high non-linearity of the corresponding models, we propose an evolutionary optimization procedure which uses intensively both conceptual and rigorous models for the design and simulation of unit operations.

Initial estimation of the total number of trays of the main distillation column, placement of feed and side streams, and steam flow rate in the simulation environment are accomplished in the conceptual design environment.

In addition, convergence of the rigorous model is enhanced by using initial estimates of internal profiles generated at the conceptual design level despite of the highly non-ideal behavior of the multicomponent azeotropic mixture. Applying the same philosophy to estimate initial values for design and operating variables of the whole process, optimal values for a cost-effective bio-ethanol purification plant are reported.

The methodology is applied to the purification process for a feed leaving the fermentation step of a conventional corn dry-grind processing facility producing 24 million liters/year of ethanol and 19 million kg/year distiller´s dry grains with solubles (DDGS). 22170 kg/hr of a mixture mainly composed by ethanol (10.80 % w/w) and water (88.98 % w/w) with traces of methanol (0.0226 % w/w) and fusel (0.2009 % w/w) are fed to the purification plant. A simplified flow diagram is shown in Figure 1.

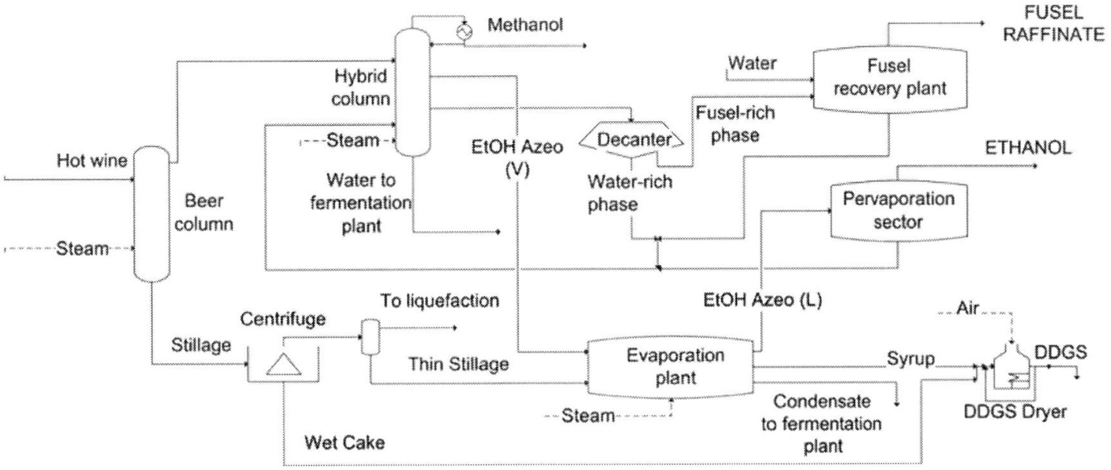

Figure 1. Simplified flow diagram of a bioethanol purification plant.

2. Initial Design

2.1. Beer Column

The ethanol and components in traces produced during the fermentation step are captured almost totally by the vapor stream leaving the stripping column. The minimum energy demand of the process (calculated as the minimum reboil ratio) is obtained using the lever arm rule by setting the bottom product as high purity water and the composition of the vapor in equilibrium with hot wine. In other words, a pinch at the top of the stripping column is considered. Figure 2 schematically shows how products of a column with an infinite number of stages move along the mass balance line as the actual reboil ratio takes values above or below the "minimum".

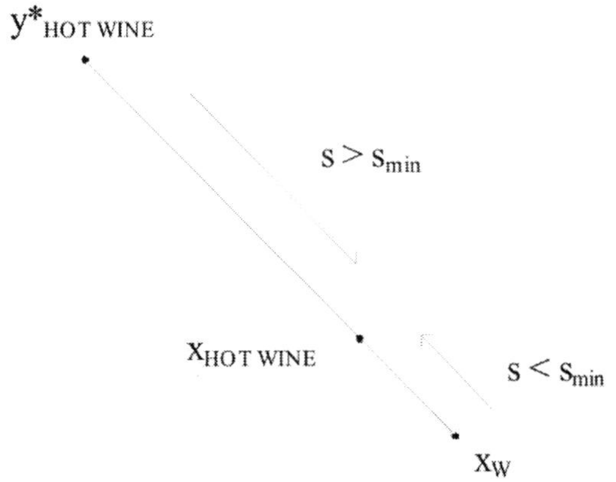

Figure 2. Influence of actual reboil ratio on the products of a column with an infinite number of separation stages.

The following steps are performed to obtain a quasi-optimal feasible design:

i) determine the maximum feasible separation and minimum energy demand (s_{min}) by applying pinch theory. An equilibrium calculation that can be performed in a conceptual model framework like DISTIL (Hyprotech, 1999) is required for such task;

ii) calculate the reboil ratio s as 1.05 s_{min};

iii) set a value for the number of stages N;

iv) simulate a stripping column with reboiler and once converged, replace the reboiler with live steam considering the reboiler duty and the latent heat of condensation of steam. This task can be done in a simulation framework like Hysys (Hyprotech, 1999).

A feasible column design was obtained, with 38 equilibrium stages, a column diameter of 0.9 m, a section pressure drop of 34.7 kPa and a steam flow rate of 3600 kg/h. The values obtained for the number of equilibrium stages and steam demand, together with the composition of ethanol in the outlet stream (51.52 % w/w) agree well with results presented in Kwiatkowski et al. (2006). Table 1 shows the "limiting" compositions of the streams entering and leaving the stripping column together with the minimum value of the reboil ratio s_{min}.

Table 1. Overall mass balance and energy demand of a beer column with and infinite number of stages (DISTIL).

	$x_{Hot\ wine}$	$y^*_{Hot\ wine}$	x_W	s_{min}
Methanol	0.000137	0.000659	0	
Ethanol	0.045288	0.303358	0	$= (x_W - x_{Hot\ wine})/(x_{Hot\ wine} - y^*_{Hot\ wine})$
Water	0.954135	0.692560	1	$= 0.1753$
1-Pentanol	0.000440	0.003423	0	

2.2. Hybrid Column

A conceptual design and rigorous simulation process is proposed for the separation of a multicomponent mixture, formed by ethanol and water with traces of methanol and fusel, in a single column. First, ethanol and methanol are lumped into one pseudo-component in order to obtain a first estimation of steam flow rate, number of stages necessary to separate an ethanol-rich stream from a fusel-rich stream, feed stage and side stream location (DISTIL). Rigorous simulation of the quaternary mixture is performed in the Hysys environment, where separation between a methanol-rich stream and an ethanol-rich stream is considered by taking into account the distillation line departing from the composition of the distillate (performed in the conceptual environment DISTIL). Finally, both columns are integrated into a single one (Hysys).

2.2.1. Side-Stream Column

For obtaining a feasible design, three steps were performed:

i) calculate $s_{min}^{(t)}$ for the separation ethanol/water/1-pentanol (DISTIL),

ii) estimate the number of equilibrium stages $N_{stages}^{(t)}$, feed stage $N_{feed}^{(t)}$ and side stream location $N_{Side\ Stream}^{(t)}$ for $s^{(t)} > s_{min}^{(t)}$ (DISTIL),

iii) simulate the quaternary system (Hysys) for the design and operating variables obtained in step ii), iv) calculate the steam flow rate $V_{steam}^{(q)}$ through the energy balance and simulate the column without reboiler (Hysys), v) simulate the system side-stream column plus decanter and water-rich phase recycle (Hysys).

Table 2 presents the results of the conceptual design performed in DISTIL[6], steps i) and ii), for the ternary mixture ethanol/water/1-pentanol, the latter used to approximate the behavior of the component fusel.

Table 2. Results for the ternary system ethanol/water/1-pentanol (DISTIL).

$R_{min}^{(t)}$ / $S_{min}^{(t)}$	2.839 / <u>1.00</u>
$R^{(t)}$ / $S^{(t)}$	2.839 / <u>1.00</u>
$N_{stages}^{(t)}$ [including condenser and reboiler]	17 [0 + 1-15 + 16]
$N_{feed}^{(t)}$/ $N_{Side\ Stream}^{(t)}$	4/11

Compositions of both ethanol and fusel in the bottom stream were selected taking into account the behavior of the residue curve (DISTIL) corresponding to the liquid in equilibrium with the vapor feed in the neighborhood of water vertex. As both for the ternary and quaternary mixture, the residue curve approaches the water vertex with compositions of ethanol above the mole fraction of fusel, the selected bottom composition reflects this behavior. The side stream is located inside the liquid-liquid gap as this stream will be separated in a decanter into a fusel-rich phase (feed to the fusel plant) and a water-rich phase (recycle to column).

Minimum reboil ratio for the given separation does not require an infinite number of stages. This behavior can be explained in terms of bifurcation of adiabatic profiles and it will be subject of further analysis in future work.

2.2.2. Methanol Column and Hybrid Column

The distillate stream of the side-stream column contains small amounts of methanol that must be separated from ethanol to agree with stringent quality standards (0.1 % V/V methanol in dehydrated ethanol at 20 °C). At a first glance, this separation could be performed in another distillation column. A different alternative would be to integrate this column with the side-stream column. The distillate line (DISTIL) corresponding to the distillate composition of the side-stream column resembles the behavior of the internal profile at total reflux and it can be considered as a good approximation to the actual operation of the methanol rectifying column, as the distillate flow rate of this column is low enough to produce a high reflux ratio. Therefore, it is possible to estimate the optimal number of stages in the rectifying section of the hybrid column as the number of trays necessary to separate the methanol in excess from the ethanol-rich stream. The methanol-rich distillate stream will also contain small amounts of ethanol and water due to the distillation line running from pure methanol to ethanol-water azeotrope.

A loss of about 0.18 % w/w of ethanol in methanol-rich distillate occurs. The side streams are located near the maximum in ethanol (ethanol-rich stream, 88.98 % w/w) and fusel (fusel-rich stream, 18.5 % w/w), respectively. The column has 35 equilibrium stages, a column diameter of 1.372 m, a section pressure drop of 11.8 kPa and a steam flow rate of 1800 kg/h. The vapor ethanol-rich stream is diverted to the first effect of the evaporation sector to provide heating while minimizing the steam demand of the plant. The condensed ethanol-rich stream is then fed to the pervaporation sector to remove the excess water.

2.2.3. Fusel Plant and Pervaporation Sector

The fusel-rich stream leaving the decanter is fed to the fusel sector where the stream is washed with water to recover about 96 % of the incoming ethanol. The resulting water-rich stream is recycled to the hybrid column. An overall amount of 363 kg/h wash-water and seven separation steps are necessary for this task. The conceptual design of a cross-flow operation is performed using DISTIL, while process simulation is done in Hysys.

A conceptual design of pervaporation (PVA/PAN MOL 1140 membrane by GFT, Germany) following the model proposed by Vier (1995) and Bausa and Marquardt (2000) was implemented in Delphi environment (Borland, 1997) to determine pseudo optimal operating values for both maximum temperature (90 °C) and permeate pressure (2.026 kPa). The model was also implemented in Hysys as a user operation extension, which calculates the required membrane area a_{req} as $a_{req} = 1.25\ a_{min}$ (Bausa and Marquardt, 2000). Both heat and refrigeration duties are calculated in Hysys from the energy balance. Behavior of trace components is taken into account by using water-alcohol separation factors. Approximate values were taken from D. Van Baelen et al. (2005). High purity bioethanol (99.5 % v/v) is obtained with an actual membrane area of 930 m^2, while the water-rich permeate is recycled to the hybrid column.

2.2.4. DDGS Plant

The co-product plant is formed by a decanter centrifuge which separates the bottom stillage from the beer column in a wet cake (35 % solids, 2683 kg/h water) and a thin stillage (18443 kg water/h). Approximately 4626 kg water/h is recycled into the second step of the liquefaction process, while 13817 kg water/h is feed to a three-effect evaporator. The resulting syrup (35 % solids, 1287 kg water/h) is mixed with the wet cake coming from the centrifuge and sent to a rotary drum dryer. While the multiple effect evaporator is simulated in Hysys, only mass and energy balances for the dryer are incorporated in Hysys. The conceptual design of the dryer is performed according to the method presented in Ulrich and Vasudevan (2004) and data from National Corn-To Ethanol Research Center (2005-2007). Table 3 summarizes the investment and operation costs of the DDGS sector. The operation cost of 35.9 $/ton DDGS agrees well with the value reported by Batelle Memorial Institute (2005).

Table 3. Operation ($/ton DDGS) and Investment ($) Costs corresponding to DDGS plant.

Item	Characteristics	Invest./Op. Cost*	Invest./Op. Cost†
Rotary Dryer	D [m]= 1.664 L [m]= 13.31 τ [min]= 19.6 rpm= 2.869 Q_{air} [kg/h]= 39150 $\eta_{Nat.\ Gas}$ = 0.048‡	1.39 10^6/19.74	1.57 10^6
Evaporator	$Area_{overall}[m^2]$= 940 Pressure[kPa]= from 10-30 kPa Q_{steam} [kg/h]= 2700	1.523 10^6/16.16	1.47 10^6
Decanter Centrifuge		1.07 10^6§/not calculated	1.07 10^6

3. Evolutionary Optimization

Once an initial design is obtained, the following improvement opportunities were tested: i) design of the beer column in the neighborhood of minimum energy demand (done from the very beginning of design process), ii) heat integration between the hybrid column and evaporation first effect (1270 kg/h of steam are saved), iii) heat recovery from the hot air leaving the dryer (saving 0.014 kg natural gas/kg water evaporated).

Finally, a change in distillate composition of the hybrid column is proposed in order to capture the trade-offs between distillation and pervaporation costs. Resorting again to the conceptual design of the hybrid column for a set of distillate compositions, results shown in Figure 3 are obtained. The decrease of investment costs for distillate compositions richer in ethanol is related to a decrease in the membrane area needed to obtain dehydrated bioethanol (a_{req}=848 m^2). Table 3 shows both investment and operating costs of the quasi-optimal plant. All cases include improvements i), ii) and iii) mentioned above. Investment

* From Ulrich & Vasudevan (2004)
† Overall operating costs = 34.55 $/ton DDGS.
‡ kg natural gas/ kg of water evaporated.
§ From Batelle Memorial Institute (2005)

and operation costs are reduced by 6.64 % with respect to the base case and by 11.48 % with respect to the worst case analyzed. Cost coefficients used to obtain the reported values are C_{steam}=2.396E-2, C_{water}=5.73E-3, C_{refr}=3.824E-2, C_{electr}=0.08 all in [\$/kWh], $C_{nat\ gas}$=289 \$/tn, $C_{memb\ repl}$ = 400 \$/m^2.

Table 4. Overall costs for a bioethanol plant producing 24 million liters/year. The facility considers both the ethanol and co-product processing plants.

	Investment, \$	Operating \$/h	Investment, \$/h	Total, \$/h
DDGS Plant	3.985 E+06	96.37	81.07	177.44
Separation	3.768 E+06	117.96	76.66	194.62
Total	7.753 E+06	214.33	157.73	372.06

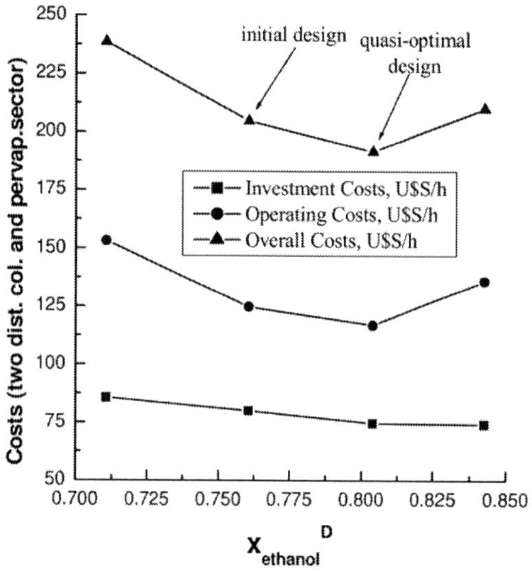

Figure 3. Overall investment and operating costs for the two distillation columns and pervaporation sector versus ethanol mole fraction in the distillate of the hybrid column.

4. Conclusions

A cost effective design for a bio-ethanol separation plant using conceptual design followed by rigorous simulation is found. The advantage of the method is that the systematic use of conceptual models allows the designer to capture the main characteristics of each operation involved in the process.

The minimum in the operation costs corresponds to a minimum in the steam flow rate of the hybrid column (1600 kg/h). The minimum in steam flow rate can be only explained by the presence of the fusel component, which influences both the energy demand and feasible products of the process. Therefore, designs based on the binary system ethanol-water do not represent the system behavior in an accurate way. From this consideration emerges the importance of properly determining the amount of trace components that enter the purification process.

Some other membranes should be considered as alternatives for the pervaporation sector in order to improve fluxes and selectivities, as it was shown that the hybrid process is economically attractive enough to promote future research efforts.

5. Acknowledgements

This work was supported by UNL, UNS, CONICET and ANPCyT from Argentina.

References

Batelle Memorial Institute, Pacific Northwest Division, 2005, Quantifying Biomass Resources for Hydrothermal Processing II.

Bausa, J. and W. Marquardt, 2000, Shortcut Design Methods for Hybrid Membrane/Distillation Processes for the Separation of Nonideal Multicomponent Mixtures, Ind. Eng. Chem. Res., 39, 1658-1672.

Borland International Inc., 1997, Scotts Valley, USA, Delphi 3 User Manual.

Hyprotech Ltd., 1999, Calgary, Canada, Hysys & Distil User Manuals.

Kwiatowski, J. R; McAloon, A. J.; Taylor, F. and D. B. Johnston, 2006, Modeling the Process and Costs of Fuel Ethanol Production by the Corn Dry-Grind Process, Industrial Crops and Products, 23, 288-296.

National Corn-To Ethanol Research Center, 2005-2007, Utilizing the National Corn-to-Ethanol Pilot Plant to Develop a Predictive Model for Distillers Dried Grain for the Fuel Ethanol and Animal Feed Industries.

Ulrich, G. D. and P. T. Vasudevan, 2004, Chemical Engineering Process Design and Economics, A practical guide. 2nd ed., Process Publishing, Durham, New Hampshire.

Van Baelen, D.; Van der Bruggen, B.; Van den Dungen, K.; Degreve, J. and C. Vandecasteele, 2005, Pervaporation of water-alcohol mixtures and acetic acid-water mixtures. Chemical Engineering Science, 60, 1583-1590.

Vier, J.,1995, Pervaporation azeotroper wässriger und rein organischer Stoffgemische-Verfahrensentwicklung und –integration, Ph. D. Thesis, Institut für Verfahrenstechnik, RWTH Aachen, Shaker Verlag, Aachen, Germany.

BIODIESEL PRODUCTION USING SUPERCRITICAL ALCOHOLS IN BATCH AND CONTINUOUS REACTORS

P. Valle[1], A. Velez[2], G.Mabe, P. Hegel[2], E.A. Brignole[2*]

[1]LEC-ICEx – DQ, Universidade Federal de Minas Gerais, Av. Antônio Carlos 6627 Belo Horizonte – MG, BRASIL
[2]PLAPIQUI, Universidad Nacional del Sur – CONICET CC 717, 8000 Bahía Blanca, ARGENTINA
**ebrignole@plapiqui.edu.ar*

Abstract. Recent experimental studies on non-catalytic transesterification have shown that high reaction rates, which justify the commercial application of this process, can be obtained if the operating conditions are above the critical pressure and temperature of the alcohol. Even though it is generally agreed that temperatures above 550 K and high alcohol/oil ratios are required to obtain high conversions and high reaction rates, there is disagreement among the different authors regarding the operating pressure, the assumed phase conditions, the use of co-solvents and the justification of the sudden increase of the rate of reaction with temperature. In the present work, the reactor phase transitions are directly observed in a double windowed cylindrical reactor and the conversion to ethyl esters is measured. Also studies on a continuous reactor are carried out to obtain a better control of operating conditions and reactants residence time. In these studies different oils, crude and refined, are studied. Also the effect of using methanol or ethanol on the rate of reaction and efficiency of conversion are considered. The optimization of the process conditions was carried out based on a statistical design of experiments where the key process variables were studied over different ranges to obtain a reliable model for the efficiency of the reaction as a function of reactants residence time, temperature, pressure and molar ratio of ethanol/oil. The present results confirm preliminary studies that indicate that high conversions can be obtained at pressures of 10-15 Mpa at temperatures between 570 and 600 K using a molar ratio of 39. From direct observations and the modeling of the phase behavior, a better understanding of the supercritical ethanol transesterification process is obtained as well as the confirmation of the phase equilibrium predictions based on the GCA-EOS model.

Keywords: Biodiesel, supercritical methanol, continuous reactor

1. Introduction

Nowadays biofuels play an important role due to the increasing worldwide energy problems. The production of methyl and ethyl esters is of great industrial interest because of their direct use as biodiesel. Most of the biodiesel currently produced uses soybean oil, methanol and an alkaline catalyst. The main products of the reaction are fatty esters and glycerin. The high value of soybean oil as a food product makes production of a cost-effective fuel very challenging. However, crude oils, animal fats and waste cooking oils can also be converted into biodiesel. Some problems with processing these low cost oils are related to the presence of water and free fatty acids. The conventional alkaline process and acid catalyst process are very sensitive to these components in vegetable oil. Most industries use pre-treatment steps to reduce the free fatty acids and water contents that hinder the efficient use of waste cooking oils, animal fats and crude oils as feedstock for biodiesel production. Besides, when acid or basic catalysts are used at moderate temperatures the reaction is slow and, in general, the removal of the catalyst and saponification products from the fatty esters + glycerin mixtures turns biodiesel and glycerin purification difficult. The transesterification of vegetable oils using supercritical alcohols is an alternative for biodiesel industrial production. The pionering experimental studies of non-catalytic transesterification by Saka and Kusdiana (2001) have shown that the process is not sensitive to both free fatty acids and water contents, and high reaction rates are observed at conditions close to the critical properties of methanol.

1.1. Phase equilibrium engineering of non-catalytic supercritical transesterification

In a previous work (Hegel et al., 2007) it was determined how the phase condition, homogeneous or heterogeneous, influences the kinetics and efficiency of the reaction of transesterification using supercritical methanol. The evolution of the phase behavior with temperature, during the non-catalytic transesterification reaction, with and without propane as a co-solvent, was observed in a see-through, two window cylindrical reactor, at controlled reaction cell global densities and initial methanol / soybean oil composition. Cell global density (ρ_g= defined as the total mass charged into the reactor divided by the reactor volume) was an important factor from a phase equilibrium engineering point of view. A high global density (ρ_g > 0.8 g/cm^3) produced a steep increase in pressure when the liquid phase completely filled the volume of the reactor. At smaller global densities (ρ_g < 0.55 g/cm^3) the liquid-vapor (LV) equilibrium conditions prevailed up to 563K. However, it was observed that high efficiencies were achieved at temperatures over 550K and 10 minutes of reaction, even in the LV phase regions [3]. This result partially explains why different operating conditions have been proposed for non – catalytic supercritical transesterification reactions.

Saka and Kusdiana (2001) used a small 5 cm^3 batch reactor, charged with rapeseed oil and methanol. Therefore, high global densities were used by the authors to study the reaction at around 620 K, resulting in very high pressures (≈45 MPa). Under these rather extreme conditions the conversion of oil to biodiesel was completed in approximately 5 minutes.

Recently Cao et al. (2005) proposed the use of propane as a co-solvent to reduce the critical point of the system and facilitate the operation in a single-phase supercritical condition. Their results indicate that it is possible to operate at milder conditions, keeping a high reaction rate, using a propane to methanol molar ratio in the range of 0.05 to 0.1, and working at 540 K and pressures around 12 MPa. The conditions proposed by these authors are quite different from those of Saka and Kusdiana (2001) and Kusdiana and Saka (2004). Hegel et al. (2007) observed in their experiments that the increase of propane concentration drastically reduces the critical temperature of the system, leading to a single supercritical phase at lower temperatures. According to them, the low pressures obtained by Cao et al. (2005) are probably also due to the use of lower cell global densities than Saka and Kusdiana (2001).

In the present work, a phase equilibrium engineering analysis, from direct observations and modeling of the phase behavior is applied to the transesterification of *Raphanus sativus L.*

oil (a non-edible vegetable oil) in supercritical ethanol, in order to find feasible homogeneous operating conditions at moderate pressures. Then, the different variables affecting the non-catalytic transesterification reaction are evaluated to set the best operating conditions. The optimum values of the variables in the experimental range studied are determined by applying a three variable Doehlert design of experiments and response surface methodology. Different vegetable oils (crude sunflower oil and commercial soybean oil) are also tested. The optimum operating conditions are being implemented in a continuous reactor.

2. Experimental

Raphanus sativus L. oil extracted by cold press, commercial soybean oil, and crude sunflower oil were used as vegetable oils. Ethanol (99.5%), methanol (99.7%) and heptane (99.%) were purchased from Anedra. Methyl heptadecanoate (> 99%) was purchased from Sigma-Aldrich.

Supercritical biodiesel reactions were carried out in batch and continuous mode. A stainless steel tubular reaction cell of 32 cm^3 (15 cm of internal diameter) was used to carry out the transesterification of crude *Raphanus sativus L.* oil in supercritical ethanol in batch mode to determine the phase behavior of the reaction. Figure 1 shows a schematic diagram of the experimental equipment. Basically, it has two adapted glass windows which allow the experimental phase behavior observation. A camera (Mitzu cyber-eye) placed in front of one of the windows records the phase behavior. Two electric resistances of 450W connected to a temperature controller (Novus N480D) were used to heat the reaction cell. It also has an external aluminum jacket to obtain a more uniform reactor heating process. A PT-100 platinum resistance thermometer placed in the external aluminum jacked measures the temperature with an error of $\pm$ 2 K. The pressure inside the reaction cell is measured by a gauge pressure for high temperature (Dynisco PG4 series). The entire system is isolated by a fiberglass mat to reduce the loss of energy and allow a better temperature control. The reactor has also a feed line to load a given amount of co-solvent.

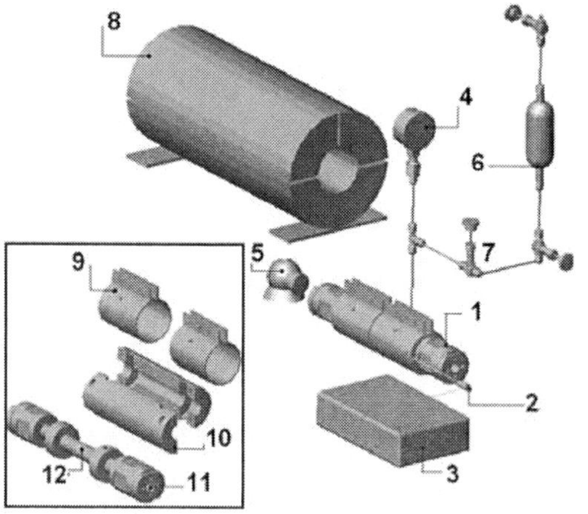

Figure 1 Equipment used in this work to analyze the phase behavior and the efficiency of a high pressure biodiesel reaction. 1. Reaction cell. 2. Temperature sensor. 3. Temperature controller. 4. Pressure sensor. 5. Camera. 6. Co-solvent reservoir. 7. Valve. 8. Fiber mat. 9. Electric resistance. 10. Aluminum jacket. 11. Windows. 12. Stainless steel reactor.

The best conditions obtained in batch mode were tested in a continuous reactor. Figure 2 shows a schematic diagram. As it can be seen the reaction system basically consists of a feed cylinder where the reactans are loaded, a membrane compressor that pumps the reactamts up to the desired pressure, a recycle line that was assembled to control the reactants flow rate, The coiled reactor of ....diameter and .... length(dimensions) is placed in an electric oven at the desired reaction temperature. The stainless steel tubular reactor has 185.64 ml internal volume (4.5 mm inside diameter), and cooling system and a back pressure regulator valve at the outlet. The temperature of the mixture is sensed by two PT-100 temperature sensors directly connected at the inlet and outlet of the reactor and the pressure is measured by a pressure gauge for high temperature at the reactor inlet.

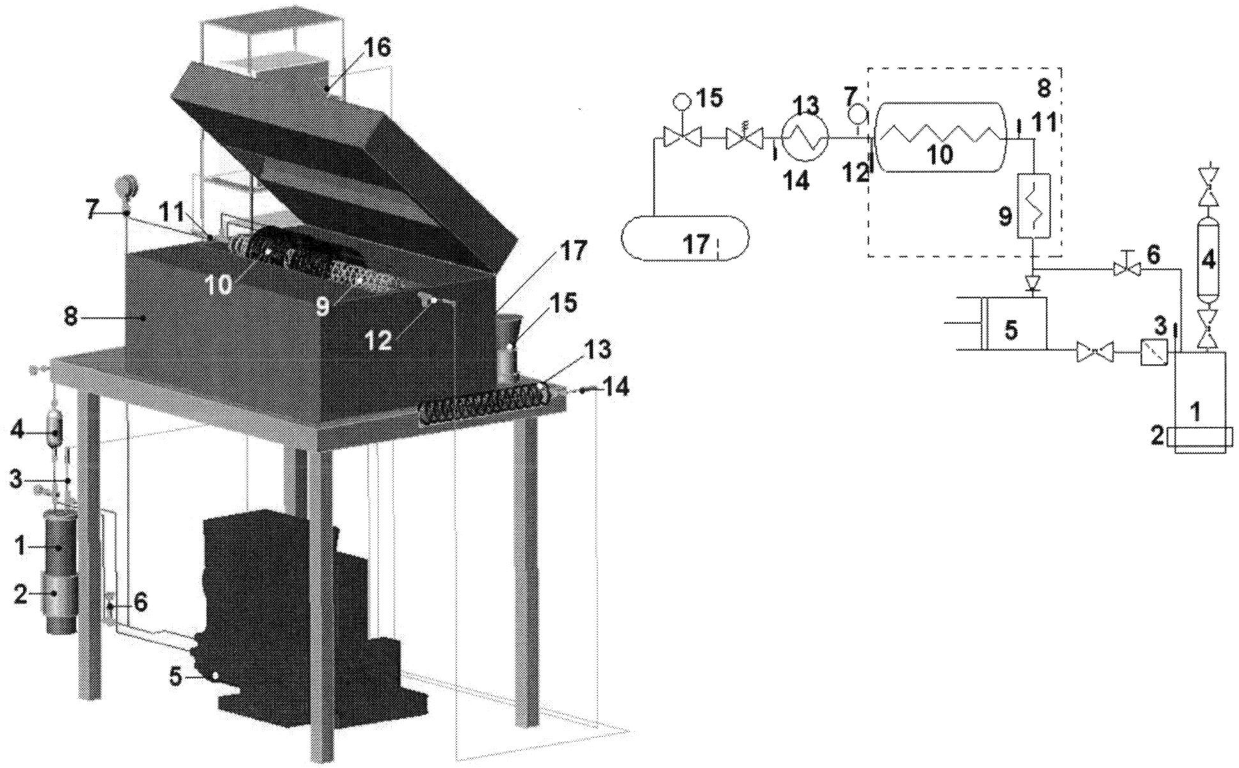

Figure 2 Continuous Transesterification reactor by supercritical alcohol 1. Suction tank. 2. Heating device. 3. Temperature sensor (PT-100). 4. CO_2 purges system. 5. Membrane compressor. 6. Needle valve. 7. Pressure gauge. 8. Oven. 9. Heating coil system. 10. Tubular reactor. 11. Temperature sensor (PT-100). 12. Temperature sensor (PT-100). 13. Cooling system. 14. Temperature sensor (PT-100). 15. Back pressure regulator. 16. Temperature Control system. 17. Discharge recipient.

2.1.Experimental Procedure

The experimental procedure of the batch mode experiences is as follows. The reactor was charged with a given amount of oil and liquid alcohol. Different molar ratios of the reagents were studied. The residual air was evacuated from the reaction cell by flushing low pressurized nitrogen gas. After the charge of reagents the entire system was assembled, and the temperature was raised up to the desired value as follows. At the beginning, the temperature controller was adjusted to give a heating rate of 15 K/min up to 493 K. Then, the system was let at 493 K for six minutes to stabilize the pressure inside. Later, the temperature controller was set to a heating rate of 10 K/min up to the reaction final temperature. The total heating time was approximately 33 minutes. The phase behavior of the system at different temperature and pressure conditions was recorded by a video camera. When the reaction finished, the temperature controller was turned off and the reactor was air-cooled to quench the reaction.

In the continuous reactor the mixture of ethanol and oil were loaded to the feed cylinder. The reactive mixture was heated at 120°C to avoid liquid partial miscibility. The membrane compressor pumped the reactants up to the desired pressure set by the back pressure regulator at the reactor outlet. A recycle line was assembled to control the reactants flow rate, which is measured volumetrically at the reactor outlet. Then the reactive mixture flow through the reactor after which was cooled at ambient temperature before being expanded at atmospheric pressure and stored for analysis. Three samples were periodically collected in a glass recipient placed at the reactor outlet after the steady state conditions was reached. After the reaction the entire system was cleaned and purged with CO_2.

2.2. Analytical method

The amount of alcohol present in the products of the reactions was separated in a vacuum oven at 70 °C and 25 mmHg. Without the presence of methanol or ethanol, the biodiesel and glycerin become practically immiscible and can be easily recovered. The biodiesel samples were analyzed by gas chromatograph (GC – Varian Star 3400 CX) to determine their ester contents. The equipment was assembled with a flame ionization detector (FID) and capillary column (J&W Scientific, model DBwax, 30 m length, 0.32 mm inner diameter, and 0.25 μm film thickness). The selected chromatograph conditions were selected according to BS EN 14103:2003. Hydrogen was used as the gas carrier at a flow rate of 1.5 ml/min, and at a split flow rate of 76 ml/min. The injector and detector temperatures were 250 °C and 270 °C, respectively. The oven temperature program consisted of starting at and maintaining 120 °C for six minutes, a ramp of 7°C/min up to 210 °C, a ramp of 15°C/min up to 250 °C, and finally maintaining 250 °C during 15 minutes. Methyl Heptadecanoate was used as internal standard. A stock solution of heptane with a known amount of internal standard was prepared (5 mg/ml). A sample solution was also prepared in heptane (17 mg/ml). The sample injected to the chromatograph consisted of 2 μL of a solution prepared with 0.2 ml of the internal standard stock solution and 0.3 ml of the sample solution (2mg/ml of methyl heptadecanoate and 10mg/ml of reaction products).

2.3. Statistical analysis

The synthesis of ethyl esters by transesterification of *Raphanus sativus L.* oil using supercritical ethanol was carried out by a Doehlert design of experiments (Doehlert, 1970). This experimental design is very useful and offers advantages in relation to other designs. Doehlert designs are easily applied to optimize variables and they need fewer experiments in relation to central composite and Box-Behnken designs (Ferreira et al., 2004).

The response selected was the ethyl esters contents in the biodiesel samples, Y_{ester}%. The variables or factors investigated were the reaction temperature, X_T, reaction time, X_t, and the initial alcohol to oil molar ratio, X_M. They were selected on the basis of a preliminary factorial design. The reagent to reactor volume ratio was fixed at 75%, in order to reach a global density of load of approximately 0.61 g/cm^3 and working pressures no higher than 18 MPa. Under these operating conditions, the pressures varied from 9,5 to 15 MPa.

The selection of levels for each variable was carried out on the basis of a previous work [3] and results obtained in preliminary studies in our lab. The lowest temperature value was set at 568 K and the highest was chosen as 598 K. The reaction time varied between 15 and 29 minutes, and the initial alcohol:oil molar ratio was studied in the 32:1 to 52:1 range. Once these levels were selected, the Doehlert desing of experiments was applied. The Doehlert matrix and the experimental conditions studied in this work are presented in Table 1. The first three columns of data show factor levels on a natural scale, and the next three show the coded factor levels in the dimensionless co-ordinate. All the runs were performed at random. Three experiments were carried out at the central point level, coded as "0", for experimental error estimation.

Tabla 1. Biodiesel production by transesterification of Raphanus sativus L. oil in supercritical ethanol. Experimental conditions studied in this work. Doehlert matrix and experimental results.

Run Number	Real Values			Coded Values			Y_{ester}(%)
	T K	t min.	RM	X_T	X_t	X_M	
1	598	22	42	1	0	0	93.08
2	590.5	29	42	0.5	0.866	0	95.51
3	590.5	24.3	52	0.5	0.289	0.8166	76.06
4	568	22	42	-1	0	0	91.70
5	575,5	15	42	-0.5	-0.866	0	72.32
6	575,5	19.7	32	-0.5	-0.289	-0.8166	91.37
7	590.5	15	42	0.5	-0.866	0	94.80
8	590.5	19.7	32	0.5	-0.289	-0.8166	92.32
9	575,5	29	42	-0.5	0.866	0	93.47
10	583	26.7	32	0	0.577	-0.8166	94.19
11	575,5	24.3	52	-0.5	0.289	0.8166	74.46
12	583	17.3	52	0	-0.577	0.8166	66.27
13	583	22	42	0	0	0	91.57
14	583	22	42	0	0	0	95.15
15	583	22	42	0	0	0	92.10

3. Results and discussion

3.1. Experimental phase behavior

The ester contents in all the experiments are also presented in Table 1. In all cases, a single-phase system at the end of the reaction was observed. Figure 3 shows a typical evolution of the phase behavior during the non-catalytic transesterification reaction, at the reported operating conditions.

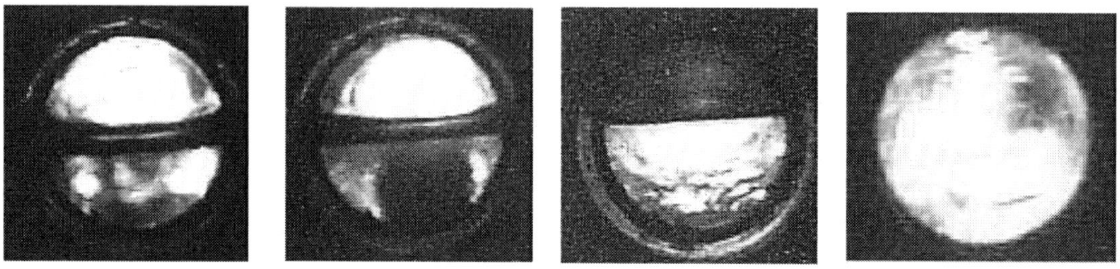

a. 298.15 K – 1 bar b. 393.15 K – 7 bar c. 493.15 K – 32 bar d. 590.5 K – 125 bar

Figure 3 Phase behavior evolution during the heating process of the reacting mixture (run #2)

Figure 3a and 3b show that the system exhibits liquid-liquid-vapor (LLV) equilibria at moderate temperatures. Figure 3c shows that at higher temperatures (such as 493 K) the liquid phases become completely miscible and liquid-vapor (LV) equilibria can be observed. However, the system changes to a single-phase condition (Figure 3d) at temperatures over 573 K, the pressure increase with temperature was moderate in all cases. According to Hegel et al. [3], this behavior is probably due to the global density of load that in this work was set around 0.615 g/cm^3.

The application of a group contribution equation with association equation of state (GCA-EOS) to the supercritical processing of fatty oil derivatives with low molecular weight solvents has been extensively discussed by Ferreira et al. (2004) and (2005) and more recently by Andreatta et al. (2007). In a previous work (Hegel et al., 2007) experimental observations of the multiple phase behavior between the reactants and products of the transesterification of vegetable oils (triglycerides, fatty esters, methanol and glycerin) were compared with modeling predictions using GCA-EOS. Phase envelope prediction in agreement with experimental trajectories of the reactive system. Andreatta et al. (2008) measured LL and LLV equilibria data for the system methyl oleate-methanol-glycerin, and used GCA-EOS to represent these experimental data. The model showed a good predictive capability. Hence, it can be applied to explore the operating conditions in the reactor. Figure 4 shows the liquid-supercritical phase behavior predicted by the GCA EOS for the methanol – methyl laureate mixture at 493 K, 523 K and 543 K in a pressure – molar concentration diagram. As it can be observed the model agrees well with the experimental data reported for the system (Shimoyama et al., 2007)

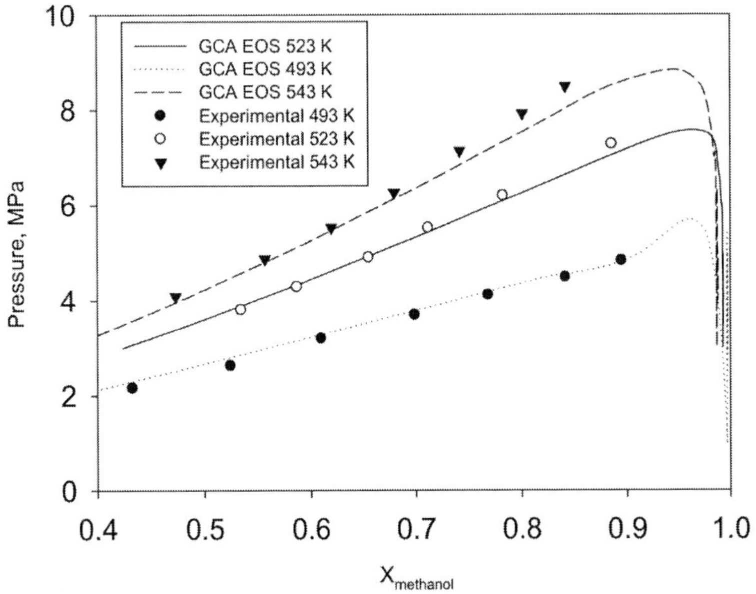

Figure 4 Liquid-vapor phase equilibria in a pressure vs. molar concentration of methanol diagram for the methanol + methyl laureate mixture. (Lines) GCA – EOS prediction, (symbols) Equilibrium data reported by Shymomaya et al.

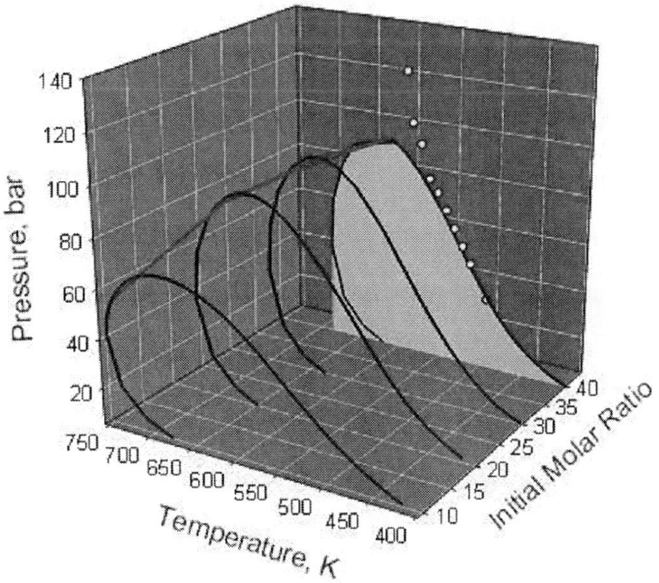

Figure 5. Phase equilibrium analysis of the transesterification of *Raphanus sativusL.* oil using supercritical ethanol. Reaction trajectory (○) on a pressure-temperature (PT) diagram for a reactive mixture with 42:1 initial EtOH:oil molar ratio. The solid line represents the GCA-EOS prediction for the liquid-vapor phase envelope for the products of the reaction (ethanol + ethyl esters + glycerin) at different initial molar ratios. The red line denotes the critical point of the reaction products.

The critical parameters of the ester (methyl laureate and ethyl eicosapentanoate) were predicted by a group contribution method (Constatinou and Gani, 1994). The GCA-EOS group energy and interaction parameters for non-associating groups are those of Espinosa et al. (2001)- For self associating groups or groups with cross-association the parameters reported by Ferreira et al. (2005) and Gros et al. (1997) are used. The alcohol, glycerin and esters interaction specific parameters are those of Hegel et al. (2008).

Figure 5 gives the phase envelopes and critical points of the system composed by ethyl oleate + ethanol + glycerin. Here, the ethyl eicosapentanoate represents the ethyl esters produced by the transesterification of *Raphanus sativus*, L oil. Phase envelopes and the critical points were predicted based on the GCA-EOS model. Figure 5 also shows the thermal history of an experimental run with the same initial ethanol:oil molar ratio (42:1). It can be observed that the predictions agree with the experimental observations of the phase conditions. In fact, a liquid single-phase condition is predicted at the end of the reaction, when almost all the vegetable oil has been converted to fatty ethyl esters, and a near supercritical state can be observed.

3.2. Model fitting and analysis of response

The data multivariable analysis is a powerful statistical technique used for studying the effect of several factors and their interactions over one or more response variables in a process (Barros et al., 2003). Response surface methodology has also been used for biodiesel production optimization (Domingos et al., 2007). In this work, response surface methodology was performed on experimental data and, as a result, a lineal regression quadratic model was obtained as shown in Eq. (1):

$$Y_{ester\%} = 92.943 + 3.729X_T + 6.369X_t - 12.465X_M - 0.553X_T^2 - 5.040X_t^2 - 14.333X_M^2 - 11.801X_TX_t + 4.572X_TX_M + 3.115X_tX_M \quad (1)$$

The error and *p-value* associated to each coefficient of this model can be seen in Table 2. According to Montgomery (2005) the *p-value* can be understood as the minimum significance level of a group of data or variable in a statistical hypothesis test.

Table 3 shows the analysis of variance – ANOVA of the lineal regression quadratic model resulted. This statistical technique subdivides the total variance of the data in groups of different sources of variance, for the purpose of statistical hypothesis tests (Barros et al., 2003). According to *p-values*, it can be observed that the model has a good adjustment with experimental data and is also statistically significant. The coefficient of determination R^2 approaches to unity (0.92) indicating that the regression model has a good prediction capacity.

Tabla 2. Coefficients of the lineal regression quadratic model and their respective error and *p-value* associated.

Coefficients		Error	*p-value*
b_0	92,943	± 1,114	0,00014
b_{XT}	3,729	± 0,965	0,06092
b_{Xt}	6,369	± 0,965	0,02220
b_{XM}	-12,465	± 0,964	0,00593
$b_{XT}2$	-0,553	± 1,762	0,78320
$b_{Xt}2$	-5,040	± 1,762	0,10356
$b_{XM}2$	-14,333	± 1,669	0,01330
$b_{XT\,.\,Xt}$	-11,801	± 2,229	0,03386
$b_{XT\,.\,XM}$	4,572	± 2,490	0,20777
$b_{Xt\,.\,XM}$	3,115	± 2,490	0,33745

Tabla 3. Analysis of variace (ANOVA) of the lineal regression quadratic model

	Degree of Freedom	Sum of Squares	Mean Squares	* Calculated value of F	*p-value*
Model	9	1258.2	139.8	6.672	0.025
Residue	5	104.8	20.95		
Lack of adjustment	3	97.3	32.44	8.708	0.105
Pure error	2	7.4	3.72		
Total	14	1363			

* The value of F is determined by the ratio between mean squares

Figure 6 plots of experimental versus predicted values for biodiesel production from *Raphanus sativus L.* oil. It can be observed that there are no tendencies in the linear regression fit; hence the model explains the experimental range studied adequately.

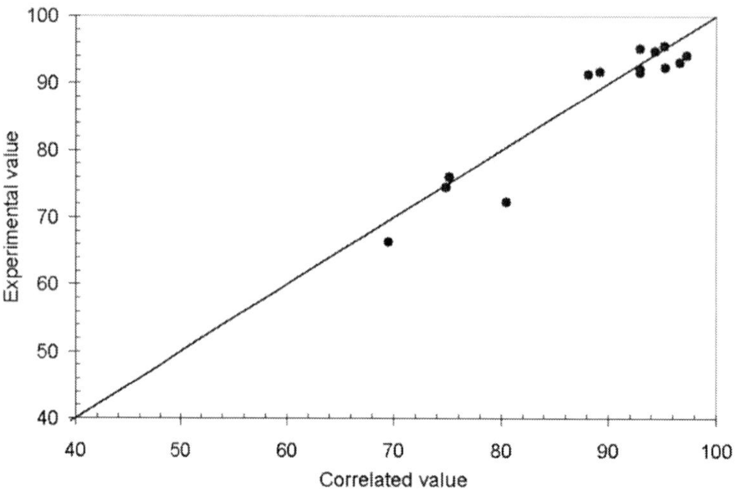

Figure 6. Raphanus sativus L. oil transesterification in supercritical ethanol. Experimental values vs. correlated values.

According to *p-values* in Table 2, from the statistical analysis, and in the experimental range studied, the EtOH:oil molar ratio is the most important factor on the transesterification of the *Raphanus sativus L.* oil, followed by time and temperature of the reaction, respectively.

To aid visualization and help to identify the type of interaction between the variables on the reaction conversion, within the studied range, the response surface for ester contents (Esters%) in biodiesel samples is shown in figure 7.

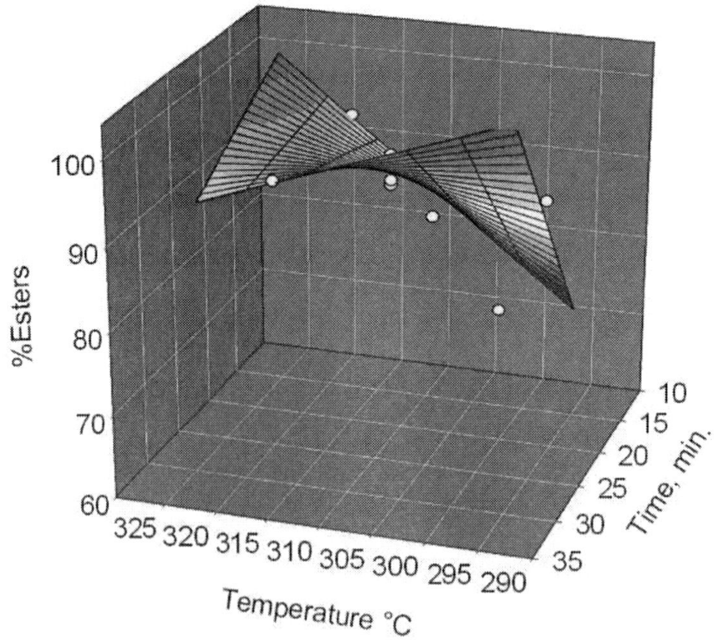

Figure 7 Biodiesel production from Raphanus sativus L. oil in supercritical ethanol. Response surface for Ester% vs. temperature for a given reaction time. EtOH:oil molar ratio 42:1. (○) Experimental data reported in table 1.

Figure 7 shows the plot of ester% versus temperature and reaction time with EtOH:oil molar ratio maintained at its zero level. According to the statistical model, there is an expected increase in ester % as temperature increases for reaction times between 15-25 minutes. For reaction times over 25 minutes, the model predicts a decrease in esters% with an increase in temperature, but we have no experimental results to confirm the observed tendency. From figure 8, it can be observed that biodiesel with ester contents higher than 95% can be obtained in 15 minutes, at 591 K.

3.3. Checking results

To check the model adequacy for predicting the maximum ester%, one additional experiment using optimal operating conditions was carried out under the conditions described in Table 4. Three replicate gas chromatography analyses indicated average contents of ethyl esters of 97.5%. The good agreement between the predicted and experimental results confirmed the validity of the model. The results derived from this study indicated that the Doehlert design is a powerful tool for optimizing the conditions for biodiesel production in supercritical alcohols.

Tabla 4. Values for testing the best conditions of maximum biodiesel production from *Raphanus sativus L. oil* in supercritical ethanol

Parameters	**Optimum value**
Esters%	97.5%
Global density of load, g/cm^3	0.61
Temperature, K	592
EtOH:oil molar ratio	39
Time, minutes	15

3.4. Transesterification of commercial soybean oil and crude sunflower oil in supercritical alcohols

Additional experimental work was carried out to test different vegetable oils. The transesterification of crude sunflower as well as soy oils was performed in supercritical ethanol using operating conditions slightly different from that reported in Table 4 (591 K, 37:1 EtOH:oil and 25 minutes). The products of these reactions were analyzed by gas chromatography in three replicates, according to AOCS Official Method Cd 11b-91, and the contents of ethyl esters in all of the chromatograms was greater than 96%. These results confirm a previous work [3], where high conversions were achieved in the synthesis of methyl ester by transesterification of soybean oil using supercritical methanol at lower temperatures (563 K) and 10 minutes of reaction time. Preliminary analysis seems to indicate that ethanol is less reactive than methanol in the experimental range considered.

3.5. High pressure continuous reactor for production of biodiesel in supercritical ethanol

In order to allow an independent and better control of pressure, temperature and residence time during biodiesel production in supercritical alcohols, the assembly proposed in Figure 6 was constructed in our lab. The best conditions obtained in this work in batch mode (reported in table 4) were applied in this continuous process. Soy oil was used to carry out the experiences with supercritical ethanol.

The results confirm previous results obtained in batch studies. For an initial molar ratio of 40:1 ethanol / soy oil and a flow rate of 3.2 g/min, 60 % of esters was obtained a 320°C and 130 bars for a reactor volume was of 93 cm^3. According to the global density of the reactive mixture measured at the operating conditions (0.431 g/cm^3) the residence time was 12 min. The global density is a key variable in the continuous process because it determines the residence time. This variable depends on the substrate that is being used in the process. In case of *Raphanus sativus L.* oil the global density for the same operating conditions was 0.61 g/cm^3.

An increased of the reactor volume to 186 cm^3 produced a biodiesel with 81.5% of ester content. Studies on continuous biodiesel production by supercritical alcohols are being carried out in our lab.

4. Conclusions

In the present work, a phase equilibrium engineering analysis, using the GCA-EOS model, was performed to explore the operating conditions in the reactor. A global density of 0.615 g/cm^3 was set to reach a single-phase system and moderate reaction pressures at the final transesterification temperatures. High ester contents in the reaction products were obtained. Design of experiments was applied to optimize the synthesis of ethyl esters from crude *Raphanus sativus L.* oil. A three variable Doehlert design has proved to be effective in the study and optimization of the variables affecting the process. A lineal regression quadratic model was obtained to predict the ester contents as a function of the variables. Analysis of the residues showed that the model predicted adequately the ester% over the experimental range considered. According to these results, in the experimental range studied, a high yield of ester (97,5%) can be obtained, working in the following conditions: approximately 592 K, EtOH:oil molar ratio of 39:1 and 15 minutes of reaction time. Different vegetable oils were tested at similar conditions and high efficient reactions were achieved. An assembly for continuous transesterification in supercritical alcohol is being prepared to confirm the best conditions obtained in bath mode reaction, as well as to reach a better and independent control of the variables that affect the reaction of transesterification in supercritical alcohols. *Raphanus sativus L.* and other oils will also be used in the continuous processes for biodiesel production.

Acknowledgments

The authors would like to thank the Argentinean National Research Council (CONICET) and CAPES (Brazilian governmental institution) for their financial support to this research, and Mr. Ivan Carlos, owner of Gênesis Sementes Armazéns Gerais Ltda, for providing the *Raphanus sativus L.* seeds used in this work.

References

Andreatta, A.E., Bottini, S., Florusse, L., Peters, C. (2007) Phase equilibria of dimethyl sulfoxide (dmso) + carbon dioxide, and dmso + carbon dioxide + water mixtures. The Journal of Supercritical Fluids. Internacional: , 42, 60.

Andreatta, A.E., Casás, L.M., Hegel, P., Bottini, S.B., Brignole, E.A. (2008) Phase equilibria in ternary mixtures of methyl oleate, glycerol and methanol. Industrial and Engineering Chemistry Research. 47, 5157.

Barros Neto, B.; Scarminio, I. S.; Bruns, R. E. (2003) Como fazer experimentos: pesquisa e desenvolvimento na ciência e na indústria.. 2 ed. Campinas: Editora da UNICAMP. 401

Cao, W., Han, H., Zhang, J. (2005) Preparation of biodiesel from soybean oil using supercritical methanol and co-solvent. Fuel, 84, 347.

Constantinou, L. and Gani, R. (1994) New contribution method for estimating properties of pure compounds, AIChE Journal, October, 1697

Doehlert, D.H. (1970) Uniform shell designs. Applied Statistics, 19, 231

Domingos, A. K. (2007) Optimization of the ethanolysis of Raphanus sativus L. crude oil applying the response surface methodology. Bioresource Technology, (article in press).

Espinosa, S. (2001) Procesamiento supercritico de productos naturales. Modelado análisis y optimización, PhD Thesis, PLAPIQUI-CONICET, Bahía Blanca, Argentina.

Ferreira, O. Brignole, E.A. and Macedo E. (2004) Modelling of phase equilibria for associating mixtures using an equation of state J.Chem.Thermodynamics 36, 1105.

Ferreira, O. Macedo, E. and Brignole, E.A. (2005) Application of the GCA-EOS model to supercritical processing of associating oil derivatives: fatty acids, alcohols and triglycerides J. Food Eng., 70, 579.

Ferreira, S. L. C. (2004) Doehlert matrix: a chemometric tool for analytical chemistry – review. Talanta, 1061

Gemma, V; Martínez, M; Aracil, J. Optimization of integrated biodiesel production. Part I: a study of the biodiesel purity and yield. Bioresource Technology, 98, 1724

Gross, H.P., Bottini, S.B., Brignole, E.A. (1997) High pressure phase equilibrium modeling of mixtures containing associating compounds and gases, Fluid Phase Equilibria, 75–87.

Hegel, P., Mabe, G., Pereda, S., Brignole, E.A. (2007) Phase transitions in a biodiesel reactor using supercritical methanol. Ind. Eng. Chem. Res. 46, 6360

Hegel, P., Andreatta, A., Pereda, S., Bottini, S., Brignole, E. (2008) High pressure phase equilibria of supercritical alcohols with triglycerides, fatty esters and cosolvents, Fluid Phase Equilibria, 31-37.

Kusdiana, D., Saka, S. (2004). Effects of water on biodiesel fuel production by supercritical methanol treatment. Bioresource Technology, 91, 289

Montgomery, D. C. Design and analysis of experiments. 6 ed. New York: John Wiley & Sons, Inc., 2005. Cap. 1, 3, 5, 6, 10 e 11.

Saka, S., Kusdiana, D. (2001) Biodiesel fuel from rapeseed oil as prepared in supercritical methanol. Fuel, 80, 25

Shimoyama, Y., Iwai, Y., Jin, B. S., Hirayama, T., Arai, Y. (2007) "Measurement and correlation of vapor-liquid equilibria for methanol + methyl laureate and methanol + methyl myristate systems near critical temperature of methanol," Fluid Phase Equilibria, 257, 217

316c The Sustainable Hydrogen Production Process

Wai Law, 2422 Shadowbrook Drive, Baton Rouge, LA 70816

With the implementation of the Hydrogen Fuel Initiative, hydrogen as an alternate energy source has drawn much interests in mitigating global warming and climate change. This research provides a framework of a closed-end-cycle for the production of hydrogen as the primary product, and the process optimization based upon the markets factors. The criteria to be met for a closed-cycle energy source are as follows: (1) to produce the carbon free based energy source, (2) with abundant supply of feedstock, (3) self-sustainable with respect to the energy consumption to the process. There are two objectives for this closed-cycle framework in addition to meeting the criteria set forth above. The first is to produce hydrogen via sea water electrolysis. The by-products for this process are the precious metals such as gold, silver and uranium recovered from the sea. The uranium is of particular interest which will in turn provides energy source to start the cycle of sea water electrolysis using energy source from the existing or the next generation nuclear reactor technology. The second objective is to resolve the world supply of fresh water problem, which in our lifetime might result in world's regional conflict as the water supply dwindling as a result of world population growth, and the environmental impact of the global economic growth. Figure 1 depicts the flow diagram for the sustainable hydrogen production processes that include the desalination process [A], the electrolysis of the desalinated water [B], the electrolysis of the brine from the desalination unit [C], the recovery of Uranium to generate electricity for the electrolysis process [D], and the using of H2 and / or nuclear process heat for the desalination process [E]. Sea water is the primary feedstock to the process and hydrogen fuel is the primary product. The operation capacity for these processes will depend on the markets factors such as the demand in electricity, hydrogen fuel, fresh water and Uranium. Research supported by the Hydrogen Fuel Initiative will improve efficiency of the electrolysis process, and decrease in cost of the electrolyzer. The recent trend of the nuclear renaissance and the generation IV nuclear technology are expected to decrease electricity costs upon the complete development of the technology. Preliminary calculation shows that the integrated hydrogen production process meets the criteria for the sustainable energy source.

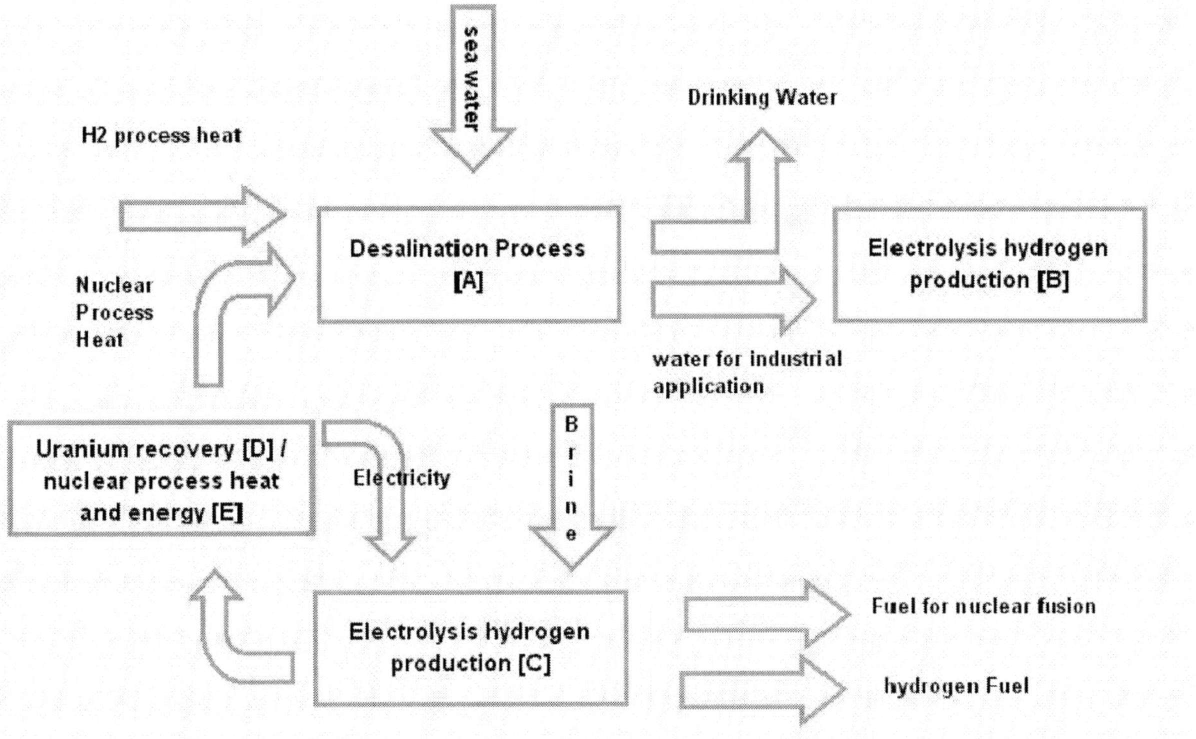

316d Toughening Vinyl Ester Resin Using Novel Bio-Based Rubber

Giuseppe R. Palmese[1], Xing Geng[1], John J. La Scala[2], and James M. Sands[3]. (1) Department of Chemical and Biological Engineering, Drexel University, 3141 Chestnut St., CAT 283, Philadelphia, PA 19104, (2) U.S. Army Research Laboratory, AMSRD-ARL--WM-MC, Building 4600, Aberdeen Proving Ground, MD 21005, (3) Army Research Laboratory, Building 4600, Aberdeen Proving Ground, MD 21005

Vinyl Ester (VE) resins are widely used to make polymer matrix composites because of their high modulus, high strength, high glass transition temperature, low weight, and low cost. However, one significant disadvantage of VE resins is low fracture toughness that limits their applications. Traditional methods using butadiene-acrylonitrile copolymer based liquid rubber as a toughener did not achieve satisfactory toughening effect due to its poor miscibility with VE resins as well as its high viscosity. In our study, a novel bio-based rubber (BR) was designed and its toughening effect on VE resins was evaluated. By using VE as a base system, it was demonstrated that significant improvement in fracture toughness (G1c=2000 J/m2 compared to 250 J/m2) is achieved with minimal reduction of Tg. Moreover, the low viscosity associated with the BR toughener provides ease of processing compared to common liquid and particulate rubber modifiers.

TRANS-FREE HYDROGENATION OF VEGETABLE FAT IN VAPOR FASE SUPERCRITICAL MODIFIED CO_2

A.Santana, M. J. Mayorga, A. Huguet, B. Pina, F. Recasens, M.A. Larrayoz*
Chemical Engineering Department, Universitat Politècnica de Catalunya.
Av. Diagonal 647, ETSEIB. Pab. G-2. 08028. Barcelona, Spain.
m.angeles.larrayoz@upc.edu

Abstract. The partial hydrogenation of fats and oils has the purpose of providing products with the desired melting profile and texture, according to their final use. The hydrogenated oil is more stable and less sensitive to oxidation. The hydrogenation of sunflower oil was carried out in fixed bed reactor in supercritical carbon dioxide – SC CO_2 (80 - 60%v/v) containing added hexane (40 - 20%v/v). The specific objective for the research presented here was the determination of supercritical carbon dioxide/hexane cosolvent mixture over a range of pressure and temperature settings to obtain an estimate of the temperature-pressure parameters that would be best for obtaining low *trans* fatty acids in the hydrogenated products. The use of a small percentage of cosolvent, may provide additional lipid solubility at lower temperature and pressure conditions and allow use of supercritical carbon dioxide/hexane at lower pressures than pure carbon dioxide, an important consideration when designing large scale processing equipment. The results from this study show that the solubility of vegetable oils in SC CO_2 can be increased due to the addition of an appropriate cosolvent, e.g. hexane, which allows a reduction in operating pressure which assures complete miscibility in the reactive mixture for a given temperature and feed composition in comparison with the system using only SC CO_2 as solvent. On the other hand, the feasibility of continuous single-phase hydrogenation increases because of the minimization the risk due to the use of modified SC CO_2 under mild conditions instead of some flammable solvents such as SC propane or DME. The process safety for the entrainers was evaluated using the DOW's Fire and Explosion Index to determine the most suitable modifier for the reaction.

1. INTRODUCTION

Environmental and human safety concerns have become determining factors in chemical engineering and process development. Currently, there is a strong emphasis on the development of more sustainable processes. Solvents that have interesting potential as environmentally alternatives to organic solvents include water, ionic liquids, fluorous phases, and supercritical or dense phase fluids. Environmental sustainability is the process of making sure current processes of interaction with the environment, are pursued with the idea of keeping the environment as pristine as possible, by avoiding a severe level of degradation, for future generations. Carbon dioxide is often promoted as a sustainable solvent, as CO_2 is non-flammable, exhibits a relatively low toxicity and is naturally abundant. However, injudicious use of carbon dioxide in a process or product can reduce rather than enhance overall sustainability.

The production of margarine and shortenings from vegetable is one typical heterogeneous catalytic hydrogenation process. Its purpose is provides products with the desired melting profile and texture, according to their final use. The hydrogenated oil is more stable and less sensitive to oxidation [1].

The low rate of reaction of the conventional process [2] as well as the rise of *cis-trans* isomerization [3] is consequence of the low solubility of hydrogen and the high mass-transfer resistance in the liquid phase. Commercial oils, which are partially hydrogenated, contain normally 30-40 % *trans*-fatty acids [4]. This isomer has shown to have similar effects as saturated fats increasing serum cholesterol levels in the blood, believed to be a major cause of heart disease [5].

The transport properties and the favourable solvent of supercritical fluids make them an adequate medium for chemical reactions and offer great opportunities for process improvement [6-7]. The application of a suitable supercritical fluid to a gas-liquid hydrogenation process can bring all reactants and products into a homogeneous fluid phase. Under these conditions the problems of hydrogen solubility and gas-liquid mass-transfer resistance are removed and reaction rates are greatly increased. The main benefits of supercritical hydrogenation are process intensification (small reactors) and improved selectivity due to the independent control of temperature, pressure, and composition of the reactants at the catalyst surface.

Tacke *et al.* [8-9], Harrod *et al.* [10-12], and Ramírez *et al.* [13-14] have been reported on the last decade, that continuous single-phase process of hydrogenation on supported Pd using supercritical propane as reaction solvent increases dramatically the rate of reaction in comparison to the traditional gas-liquid process as well as reduces the *trans* fatty acid content (less than 3 wt%) in the final hydrogenated product.

Despite of the great potential that exhibits the use of SC propane or DME in an unique reactive single-phase as alternative process to obtain healthier raw materials for foodstuffs, the idea of using high flammable solvents in the presence of hydrogen under high pressure conditions leads to increase the risk of the process.

In the case of CO_2, there are many significant advantages as it is naturally occurring, inexpensive, colourless, odourless, tasteless, non-flammable, non-ozone depleting and nontoxic. It is by far; the most widely used SCF [15]. It has been employed in the hydrogenation of a wide variety of substrates where the reactive system was operated at two-phase conditions [16].

In the hydrogenation of fats and oils, CO_2 is not commonly used due to the low solubility of triglycerides in it. This fact leads to operate under two-phase conditions because, from an industrial point of view, the high pressure requirements to form a single-phase system become the process too expensive.

The process with SC propane or DME as solvent presents completely miscibility with triglycerides [17], which allow operating in a unique reactive gas phase into the reactor under mild operating conditions but Bonanza and Recasens [18] found by means of the evaluation of the DOW's Fire and Explosion Index (FEI) [19].

The safety and loss prevention guide developed by the Dow Chemical Company gives a method for evaluating the potential hazards of a process and assessing the safety and loss prevention measures needed. A numerical Fire and Explosion Index is calculated, based in the nature of the process and the properties of the materials. The larger the value of the index, the more hazardous is the process [20].

The effect of several entrainers, allowed to foodstuff processing (e.g. ethanol, propane, butane, acetone, etc.), on the continuous vegetable oil hydrogenation in SC CO_2 was theoretically studied in order to increase the process feasibility.

The selection of the possible modifiers to add to SC CO_2 was made in accordance with the EC Council directive [21] which states the regulations regarding the use of solvents in the production of foodstuffs and food ingredients within the CE.

Table 1. Extraction solvents which use are the product or fractionation of fats and oils [21].

Solvent	T_c(K)	P_c(Mpa)	ω	Dipole Moment (D)
Hexane	508	3.00	0.305	0
Ethylmethylketone	536	4.15	0.324	2.76

The use of a small percentage of cosolvent may provide additional lipid solubility at lower temperature and pressure conditions, allowing use of supercritical carbon dioxide/hexane at lower pressures than pure carbon dioxide. Among the cosolvents under study, hexane appears to be the most appropriate. The hexane was chosen as the cosolvent because it has not reacted with hydrogen compared with Ethylmethylketone and the hexane is the less reactive solvent due to fact that it is a hydrocarbon (See table 1).

The addition of an entrainer to CO_2 dramatically increases the solubility of triglycerides in the reaction medium. This fact would offer the possibility to hydrogenate in a unique-single phase into the reactor under mild operating conditions in comparison to those with pure CO_2. The results obtained from this preliminary study showed that supercritical CO_2 hydrogenation in presence 20% v/v hexane presented low conversion with low *trans* and low stearic.

2. HYDROGENATION REACTOR SETUP

2.1 Raw Materials

A sunflower seed oil from *Helianthus agnus* from Sigma Aldrich (Barcelona, Spain) that had an initial iodine value (IV) of 133 and a fatty acid composition [10,68% saturated (C16:0, C18:0), 22.23% cis C18:1, 64.65% C18:2 and 0.8% C18:3] was used in all experiments. The mixture dioxide carbon/hexane (80:20 v/v) used in this work was supplied by Abello Linde S.A. (Barcelona, Spain). The hydrogen (99.999% minimum purity grade) was purchased from Praxair (Barcelona, Spain).

The catalyst used was 0.5% Pd on alumina (spheres of 2 mm, surface area = 320m2/g, pore volume = 0.45 cm3/g and bulk density = 750 kg/m3. The metal is deposited on an eggshell on the pellets) from Johnson Matthey (Barcelona, Spain).

2.2 Equipment

Hydrogenation runs were carried out in a microreactor (Eurothechnica, 2007), the maximum pressure supported is 500 bar and 250ºC of temperature. The reactor material is titanium. The experimental reactor setup is shown in Figure 1.

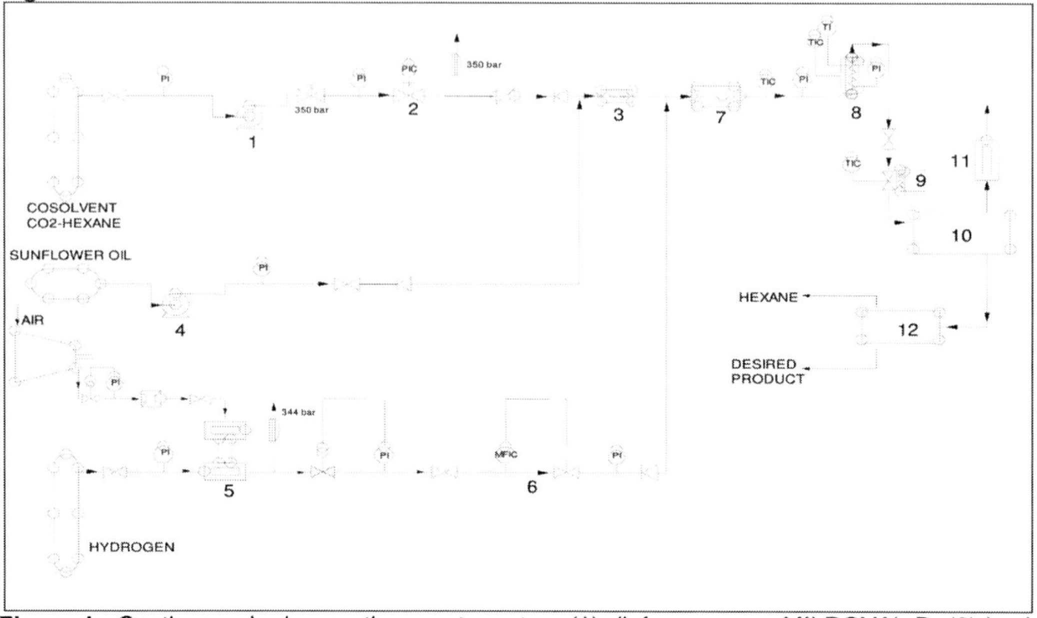

Figure 1. Continuous hydrogenation reactor setup: (1) diafragm pump MILROYAL D; (2) back pressure; (3) static mixer; (4) piston pump; (5) HASKEL gas booster AG-62; (6) Brooks Smart 5850-S; (7) heating device; (8) reactor HPA 500; (9) autoclave engineers needle-valve; (10) ethylene glycol bath; (11) rotameter; (12) Rota-vapor.

Liquefied dioxide carbon and hexane were pumped using a high pressure diaphragm pump (Milroyal D, Dosapro Milton Roy, France) to the reactor in order to provide and maintain a system downstream pressure of 18–25 MPa, which was manually set with the high pressure regulator (Model PR57, GO, Euroval, Barcelona). The sunflower oil was pumped at constant flow rate using a HPLC pump (Gilson 305, France) and H_2 was compressed by a gas booster system (AG-62, Haskel, Barcelona) equipped with a high-pressure gas reservoir. H_2 flow was metered from the constant pressure reservoir through a mass-flow indicating-controller (Model 5850S, Brooks Instruments, Euroval, Barcelona).

The oil substrate was mixed with dioxide carbon/hexane in a 20-cm long, 1/4" OD, static mixer (37-04-065, Kenics, Barcelona). H_2 was added downstream of mixer. The reactant mixture was preheated to the desired operating temperature before entering the reactor. The reactor was heated with an electrical heating jacket. Control thermocouples were located in the outside skin of the reactor body. Internal reactor temperature was monitored with a thermowell located in the bottom.

After leaving the reactor, the effluent was continuously expanded to atmospheric pressure on an externally heated needle valve in order to control the total flow of the reactor mixture.

This effluent was then sent to a series of glass U-tubes, immersed in an ethylene glycol bath held at 224 K to condense the oil from the hexane.

The molar feed concentration (oil:H_2:CO_2+hexano) was 1:4:95 mol% and the mixture of supercritical carbon dioxide/hexane was (80:20 and 60:40 in volume ratio) respectively.

2.3 Fatty acid content of the fully hydrogenated fats

The fatty acid methyl esters (FAMEs) were prepared by the method described in the French Standard (NF T 60-233, 1977). FAMEs were analyzed by capillary gas chromatography (Shimadzu GC-2010, Barcelona, Spain). Fused silica capillary column 0.25 mm X 60 m (Restek Corporation, PA, USA) was used. The injector and detector temperatures were kept at 220 ºC and 250 ºC respectively. The initial temperature of the column was 180 ºC and it was programmed to increase at the rate of 5 ºC/min to 250 ºC.

The calculated iodine value (IV) was determined simply by multiplying the percentage of each unsaturated fatty acid by its constant and addition of the results.

3. PHASE EQUILIBRIUM ESTIMATION

The addition of a modifier to CO_2 increases the single-phase region for the reaction due to the improvement of triglycerides solubility in SC CO_2. Therefore, the temperature and pressure dependence of the solvent power are enhanced. This fact makes possible to operate under milder conditions (e.g. lower pressures 20- 30 MPa) in comparison to those with pure CO_2 (75 Mpa).

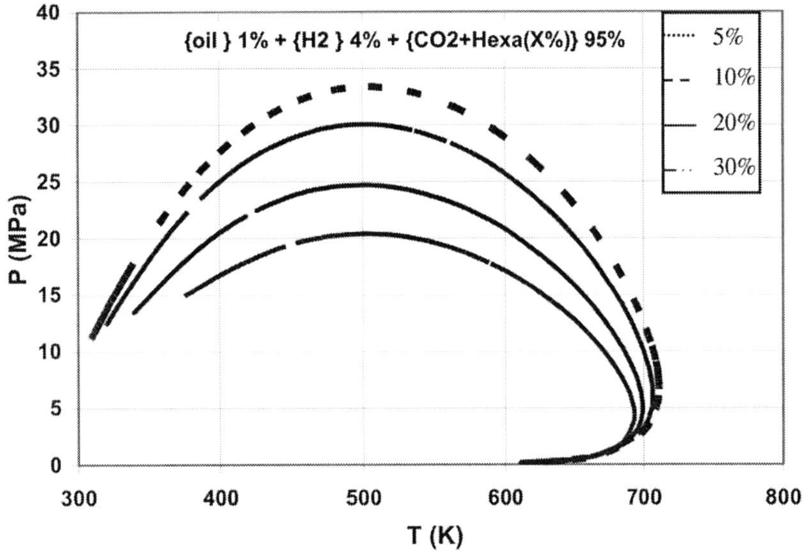

Figure 2. PT-envelope at constant mol composition (1%oil:4%H_2:95%CO_2+hexane) for different composition of hexane

To investigate the influence of modifier (hexane) on phase behaviour of the reactive mixture Sunflower oil/SC CO_2/H_2 for a defined composition, the phase envelopes were modelled using the standard Peng-Robinson equation of state (PR-EOS) with one *kij* binary parameter.

The mixture composition was set at typical values for the reaction under supercritical conditions: The molar oil and H2 concentration were kept constant at 1 and 4 mol %, respectively. Though high substrate loadings and low concentration of hydrogen are desirable.. The entrainer composition in the mixture was set at 20 % hexane of the CO2 mol concentration because higher values can make the process potentially risky due to the inherent flammability character of the most of the substances selected as potential modifiers. PT-diagrams were generated by means of ASPEN PLUS® simulation software [22]. Each isopleth plotted in these diagrams represents the phase limit between the two-phase and the one-phase regions. The temperature and pressure dependence of the solvent power are enhanced (See figure 2). We have chosen the composition 20% v/v hexane due to the conditions of equipments. The operating conditions were between 250 – 300 bars and.

4. RESULTS

Vegetable oils are considered slightly polar molecules despite of the fact that they mainly consist of mixtures of triglycerides with long-fatty acid chains [23]. The possible explanation for the larger cosolvent effect of hexane in comparison to that with ethanol is that vegetable oil is a molecule dominated by its large hydrocarbon structure rather than its polarity. The better behaviour of hexane in comparison with the other alkanes under study is explained by the fact of the solubility of the solute increases with chain length for the straight-chain of non-polar cosolvents [24].

The critical properties of the solvent and the the mixture solvent + oil are presented in Table 2 and 3.

Table 2. Critical Properties of the solvent

solvent	Tc(°C)	Pc (bar)	Vct (m^3)
80% CO_2 - 20% Hexane	132,03	123,3	0,0007
60% CO_2 - 40% Hexane	172,42	81,59	0,0010

Table 3. Critical Properties of the mixture

Mixture	Tc(°C)	Pc (bar)
Solvent (80-20)+H2+sunflower oil	183,53	251,23
Solvent (60-40)+H2+sunflower oil	211,88	137,96

The supercritical hydrogenation was carried out with a molar feed concentration (oil:H_2:CO_2+hexano) was 1:4:95 mol% and the mixture of supercritical carbon dioxide/hexane was (80:20/60:40 in volume ratio) respectively. The operating variables are shown in Table 4.

The results obtained from this preliminary study showed that CO_2 supercritical hydrogenation in presence 20% v/v hexane presented low conversion with low *trans* and low stearic (see Table 5). The final product can have suitable plastic properties. This study showed that it is possible to determine the optimum operating conditions to obtain a final hydrogenated product of immediate food application (such as for margarine/shortenings production) such that it has a low trans content (> 3%) and a moderate IV reduction together with a relatively low stearic content.

Table 4. Operating of sunflower oil hydrogenated on 2% Pd/C (1g), P=250 bar, total volumetric flow= 0.28 Kg/h and supercritical CO_2+hexane as medium

Sample	Temperature
A	200
B	185
C*	120
D*	100
E*	60

Table 5. Fatty acid composition (as weight percentage) of hydrogenated products

Sample	C18:0%	*trans* C18:1%	C18:1%	IV
A	6.4	2.4	24.77	123
B	7.5	4.0	25.17	117
C*	32.3	19.7	22.01	66.6
D*	26.7	17.2	24.70	77.13
E*	10.61	5.12	31.70	112

*Near critical CO_2+hexane

It was not possible to analyze the supercritical hydrogenation in presence 40% v/v hexane due to the pressure did not remain constant. A possible explanation would be an increase of the viscosity of the mixture. The viscosity of 20% v/v hexane is $1.5.10^{-4}$ and the viscosity of 40% v/v hexane is 2.10^{-4}.

The Table 5 shows the results obtained under supercritical fluids and near critical fluids. The Table 4 illustrates the operating variables for each sample, where the pressure remained constant and the temperature was varied. The sample A and B under supercritical conditions presented low conversion and low trans content. The experiment C, D and E were carried out under near critical conditions. The hydrogenation under low temperature resulted in a reduction of isomer *trans* and stearic contents but the oleic content was higher. No reaction was observed at temperature lower than 60ºC.

5. CONCLUSIONS

A small amount of a co-solvent increases the ability of supercritical carbon dioxide to dissolve polar compounds. Near supercritical CO_2 has dissolving properties similar to hexane. This means that, by itself, carbon dioxide is very good for dissolving relatively non-polar materials. The addition of just a small quantity of co-solvent enhances the solubilizing power of the supercritical carbon dioxide making it possible to extract much more polar molecules.

The results from this study show that the solubility of vegetable oils in SC CO_2 can be increased due to the addition of an appropriate cosolvent e.g. hexane which allows a reduction in operating pressure which assures complete miscibility in the reactive mixture for a given temperature and feed composition in comparison with the system using only SC CO_2 as solvent. On the other hand, the feasibility of continuous single-phase hydrogenation increases because of the minimization the risk due to the use of modified SC CO_2 under mild conditions instead of some flammable solvents such as SC propane or DME.

In this work we report in the fluid-phase hydrogenation on sunflower oil in supercritical dioxide carbon/hexane with 1 mol % vegetable oil, 4% H_2, and 95 % CO_2+hexane. The reaction is run continuously in a single-fluid phase, using a laboratory setup. The catalyst is 0.5% Pd/Al_2O_3.

A wide range of hydrogenation products can be obtained with certain plastic characteristics for further food application by tuning the reaction conditions. The experimental results show that one principal advantage of using supercritical carbon dioxide in presence of the cosolvent hexane is the low *trans* acid content.

6. ACKNOWLEDGMENTS

A fellowship to A. Santana from the FPI program (Spanish Ministry of Science and Technology) is acknowledged. The project of CiCYT-FEDER (Madrid,Spain) Project Number AGL 2006-05156 provided funds for a fellowship to M. J. Mayorga, A. Huguet and B.Pina. Financial support to this work was received from the same research project.

7. REFERENCES

[1] Rase, H. F., Handbook of Commercial Catalysts: Heterogeneous Catalysts, CRC Press, Boca Raton, FL (2000).

[2] Farrauto, R. J. and C. H. Bartholomew, Fundamentals of industrial catalytic processes, Chapman & Hall, Great Britain (1997).

[3] Horiuti, J., and M. Polanyi. "Exchange reactions of hydrogen on metallic catalysts", Trans. Faraday Soc., 30, 1164 (1934).

[4] King, J. W., R. L. Holliday, G. R. List and J. M. Snyder, "Hydrogenation of vegetable oils using mixtures of supercritical carbon dioxide and hydrogen", JAOCS, 78 (2), 107 (2001).

[5] Oomen, C. M., and M. C. Ocke, "Association between *trans* fatty acid intake and 10-year risk of coronary heart disease in the Zutphen eldery study: A prospective populationbased study", Lancet, 357, 746 (2001).

[6] Baiker, A, "Supercritical Fluids in Heterogeneous Catalysis", *Chem. Rev.*, 99, 453 (1999).

[7] Savage, P. E., S. Gopalan, T. I. Mizan, C. J. Martino, and E. E. Brock, "Reactions at Supercritical Conditions: Applications and Fundamentals", AIChE J., 41, 1723 (1995).

[8] Tacke, T., S. Wieland, and P. Panster, "Hardening of Fats and Oils", Proceedings of The 3rd International Symposium on High-Pressure Chemical Engineering, Zurich, Switzerland, p.17 (1996).

[9] Tacke, T., Wieland, S. and Panster, P. Selective and Complete Hydrogenation of Vegetable Oils and Free Fatty Acids in Supercritical Fluids, in

Green Chemistry Using Liquid and Supercritical Carbon Dioxide, ed. J. M. De Simone, Oxford University Press, Oxford (2003).
[10] Harrod, M., M. Macher, S. Van den Hark, and P. Moller, "Hydrogenation at supercritical conditions" Proceedings of Fifth Conference on Supercritical Fluids and their Applications, Garda, Italy, p.319 (1999).
[11] Härrod, M., S. Van den Hark, M.-B. Macher and P. Moller, Hydrogenation at supercriticalsingle-phase conditions, In High Pressure Process Technology: Fundamentals and Applications, Eds. Bertucco and Vetter, Elsevier, New York (2001).
[12] Härröd, M., Van den Hark, S., "Hydrogenation of oleochemicals at supercritical at supercritical single-phase conditions: influence of hydrogen and substrate concentrations on the process", Appl. Catal. A: Gen., 210, 207 (2001a).
[13] Ramírez, E., F. Recasens, M. Fernández, and M. A. Larrayoz, "Hydrogenation of Sunflower Oil on Pd/C in Supercritical Propane: Operating Conditions in a Continuous Internal Recycle Reactor", AIChE Journal, 50, 6 (2004).
[14] Ramírez, E., M. A. Larrayoz and F. Recasens, "Intraparticle Diffusion Mechanisms in Supercritical Sunflower Oil Hydrogenation on Pd/C Catalyst", AIChE Journal, 52, 4, 1539-1553 (2006).
[15] Jessop, P. G. and W. Leitner, Chemical Synthesis Using Supercritical Fluids, Wiley-VCH, Weinheim (1999).
[16] Hitzler, M.G., F. R. Smail, S. K. Ross, and M. Poliakoff, "Selective Catalytic Hydrogenation of Organic Compounds in Supercritical Fluids as a Continuous Process", *Org.Process Res. Dev.*, 3(2), 137 (1998).
[17] Peters, C. J., H. J. van der Kooi, J. L. de Roo, J. de Swaan Arons, J. S. Gallagher, and J.M. H. Levelt Sengers, "The search for tricritically in binary mixtures of near-critical propane and normal paraffins", Fluid Phase Equilibria,51, 339 (1989).
[18] Bonanza, N. and Recasens, F. (2005). In Proceedings of The 10th Meeting on Supercritical Fluids: Reactions, Materials and Natural Product Processing, Strasbourg/Colmar, France, p.215.
[19] Dow's Fire and Explosion Index Hazard Classification Guide. 7th ed. The Dow Chemical Company, American Institute of Chemical Engineers (1994).
[20] Coulson, J. M. and J. F. Richarson, Chemical Engineering Vol. 6, Pergamon, Great Britain (1983).
[21] EC, EC directive 84/344/EEC (1984).
[22] Aspen Plus® Version 2004.1 (2006).
[23] Fornari, T., S. B. Bottini, and E. A. Brignole, Application of a Group Contribution Equation of State to Mixture of Supercritical Gases with Natural Fatty Oils and Derivatives, Latin American App. Res., 31, 287 (2001).
[24] Dobbs, J.M., J.M. Wong and K. P. Johston, Nonpolar Co-Solvents for Solubility Enhancement in Supercritical Fluid Carbon Dioxide, J. Chem. Eng. Data, Vol. 31, 303 (1986)

316f A Preliminary Physicochemical, Exergetic and Economic Study of Hydrogen Production Utilizing Glycerol

Antonio Carlos Caetano Souza, Energy, UNESP (São Paulo State University), Av Ariberto Pereira da Cunha, 333, Guaratinguetá, 12514-410, Brazil

This work has as objective to demonstrate technical and economic viability of hydrogen production utilizing glycerol.

The volume of this substance, which was initially produced by synthetic ways (from oil-derived products), has increased dramatically due mainly to biodiesel production through transesterification process which has glycerol as main residue. The surplus amount of glycerol has been generally utilized to feed poultry or as fuel in boilers, beyond other applications such as production of soaps, chemical products for food industry, explosives, and others. The difficulty to allocate this additional amount of glycerol has become it in an enormous environment problem, in contrary to the objective of biodiesel chain, which is to diminish environmental impact substituting oil and its derivatives, which release more emissions than biofuels, do not contribute to CO2-cycle and are not renewable sources.

Beyond to utilize glycerol in combustion processes, this material could be utilized for hydrogen production. However, a small quantity of works (theoretical and experimental) and reports concerning this theme could be encountered. Firstly, the produced glycerol must be purified since non-reacted amounts of materials, inclusively catalysts, contribute to deactivate catalysts utilized in hydrogen production processes. The volume of non-reacted reactants and non-utilized catalysts during transesterification process could be reutilized.

Various technologies of thermochemical generation of hydrogen that utilizes glycerol (and other fuels) were evaluated and the greatest performances and their conditions are encountered as soon as the most efficient technology of hydrogen production.

Firstly, a physicochemical analysis must be performed. This step has as objective to evaluate the necessary amount of reactants to produce a determined volume of hydrogen and determine thermodynamic conditions (such as temperature and pressure) where the major performances of hydrogen production could be encountered. The calculations are based on the process where advance degrees are found and hence, fractions of products (especially hydrogen, however, CO2, CO, CH4 and solid carbon could be also encountered) are calculated. To produce 1 Nm3/h of gaseous hydrogen (necessary for a PEMFC - Proton Exchange Membrane Fuel Cell - containing an electric efficiency of about 40%, to generate 1 kWh), 0,558 kg/h of glycerol is necessary in global steam reforming, 0,978 kg/h of glycerol in partial oxidation and cracking processes, and 0,782 kg/h of glycerol in autothermal reforming process. The dry reforming process could not be performed to produce hydrogen utilizing glycerol, in contrary to the utilization of methane, ethanol, and other hydrocarbons.

In this study, steam reforming process was preferred due mainly to higher efficiencies of production and the need of minor amount of glycerol as cited above. In the global steam reforming of glycerine, for one mole of glycerol, three moles of water are necessary to produce three moles of CO_2 and seven moles of H_2.

The response reactions process was utilized to predict steam reforming process more accurately. In this mean, the production of solid carbon, CO, and CH_4, beyond CO_2 and hydrogen was predicted. However, traces of acetaldehyde (C_2H_2), ethylene (C_2H_4), ethylene glycol, acetone, and others were encountered in some experimental studies. The rates of determined products obviously depend on the adopted catalysts (and its physical and chemical properties) and thermodynamic conditions of hydrogen production. Eight reactions of steam reforming and cracking were predicted considering only the determined products.

In the case of steam reforming at 600°C, the advance degree of this reactor could attain its maximum value, i.e., overall volume of reactants could be obtained whether this reaction is maintained at 1 atm. As soon as temperature of this reaction increases the advance degree also increase, in contrary to the pressure, where advance degree decrease as soon as pressure increase. The fact of temperature of reforming is relatively small, lower costs of installation could be attained, especially cheaper thermocouples and smaller amount of thermo insulators and materials for its assembling.

Utilizing the response reactions process in steam reforming, the predicted volumes of products, for the production of 1 Nm^3/h of H_2 and thermodynamic conditions as cited previously, were 0,264 kg/h of CO (13% of molar fraction of reaction products), 0,038 kg/h of CH_4 (3% of molar fraction), 0,028 kg/h of C (3% of molar fraction), and 0,623 kg/h of CO_2 (20% of molar fraction). Through process of water-gas shift reactions (WGSR) an additional amount of hydrogen could be produced utilizing mainly the volumes of produced CO and CH_4. The overall results (steam reforming plus WGSR) could be similar to global steam reforming. An attention must to be taking into account due to the possibility to produce an additional amount of CH_4 (through methanation process) and solid carbon (through Boudouard process). The production of solid carbon must to be avoided because this reactant diminishes (filling the pores) and even deactivate active area of catalysts. To avoid solid carbon production, an additional amount of water is suggested. This method could be also utilized to diminish the volume of CO (through WGSR process) since this product is prejudicial for the activity of low temperature fuel cells (such as PEMFC). In some works, more three or even six moles of water are suggested.

A net energy balance of studied hydrogen production processes (at 1 atm only) was developed. In this balance, low heat value of reactant and products and utilized energy for the process (heat supply) were cited. In the case of steam reforming utilizing response reactions, global steam reforming, and cracking processes, the maximum net energy was detected at 700°C. Partial oxidation and autothermal reforming obtained negative net energy in all cited temperatures despite to be exothermic reactions. For global steam reforming, the major value was 114 kJ/h. In the case of steam reforming, the highest value of net energy was detected in this temperature (-170 kJ/h). The major values were detected in the cracking process (up to 2586 kJ/h).

The exergetic analysis has as objective, associated with physicochemical analysis, to determine conditions where reactions could be performed at higher efficiencies with lower losses. This study was performed through calculations of exergetic and rational efficiencies, and irreversibilities. In this analysis, as in the previously performed physicochemical analysis, conditions such as temperature of 600°C and pressure of 1 atm for global steam reforming process were suggested due to lower irreversibility and higher efficiencies. Subsequently, higher irreversibilities and lower efficiencies were detected in autothermal reforming, partial oxidation and cracking process. Comparing global reaction of steam reforming with more-accurate steam reforming, it was verified that efficiencies were diminished and irreversibilities were increased. These results could be altered with introduction of WGSR process.

An economic analysis could be performed to evaluate the cost of generated hydrogen and determine means to diminish the costs. This analysis suggests an annual period of operation between 5000-7000 hours, interest rates of up to 20% per annum (considering Brazilian conditions), and pay-back of up to 20 years. Another considerations must to be take into account such as tariffs of utilized glycerol and electricity (to be utilized as heat source and (or) for own process as pumps, lamps, valves, and other devices), installation (estimated as US$ 15.000 for a plant of 1 Nm3/h) and maintenance cost. The adoption of emission trading schemes such as carbon credits could be performed since this is a process with potential of mitigates environment impact. Not considering credit carbons, the minor cost of calculated H2 was 0,16288 US$/kWh if glycerol is also utilized as heat sources and 0,17677 US$/kWh if electricity is utilized as heat sources. The range of considered tariff of glycerol was 0-0,1 US$/kWh (taking as basis LHV of H2) and the tariff of electricity is US$ 0,0867 US$/kWh, with demand cost of 12,49 US$/kW. The costs of electricity were obtained by "Companhia Bandeirante", localized in São Paulo State. The differences among costs of hydrogen production utilizing glycerol and electricity as heat source was in a range between 0,3-5,8%.

This technology in this moment is not mature. However, it allows the employment generation with the additional utilization of glycerol, especially with plants associated with biodiesel plants. The produced hydrogen and electricity could be utilized in own process, increasing its final performance.

Superficial and structural behaviour of activated carbon, obtained from bamboo, as adsorbent of mercury ions

Pedro González García [a,b] , *Yolanda Pliego Bravo* [b]

Departamento de Química Inorgánica, Facultad de Ciencias Químicas, Universidad Complutense de Madrid, Avenida Complutense s/n, 28040 Madrid, Spain

División de Estudios de Posgrado e Investigación, Instituto Tecnológico de Orizaba, Av. Oriente 9 No. 852, 94320 Orizaba, México

Abstract

Activated carbon was preparing by means of design of experiments starting from three bamboo species. The results show wide area superficial values (280 – 1500 m^2/g), pH average of 9.91, yields from 20 to 85 % and high content of functional superficial oxide acid groups. Scanning electron microscopy analysis suggest the evolution of the porosity in the material as result of the activation time where is possible to observe the increase in the radii of the porous channels. Adsorption studies by Freundlich isotherm indicates the formation of single layer, the kinetic behavior is described by Fleming's empirical model and diffusion studies indicates the superficial diffusion as controlling stage.

Introduction

Due to activated carbon (AC) is a widely used material[1,2,3], starting materials to obtain it must have enough carbon content such as wood, tar, coconut shells, fruit's nuts, anthracite and more. AC production mainly involves the next stages: raw material preparation, carbonization and activation. Physical activation, usually includes carbonization and activation in combined processes. Activation was done starting 400 ^{o}C in a partial gasification with an oxidant agent (water steam, air, CO_2 and oxygen) with the purpose of developing porosity and an expected area superficial. At this conditions steam water reacts creating CO and gas hydrogen, described by the next reaction[4]:

$$C_{(s)} + H_2O \longrightarrow CO_{(g)} + H_{2(g)}$$

Depending of reaction conditions between raw material, activation atmosphere, temperature, oxidant agent and even raw material presentation, functional superficial oxygenated acid groups (FSOAG) are created on the AC surface. Bamboo in Mexico , has been designated mainly to the furniture elaboration and rural constructions nearby to the place where the bamboo species grow[5].

The purpose of this work is the use of bamboo species *Guadua Angustifolia, Bambusa Vulgaris Striata* and *Bambusa Oldhammi*, in the production of AC by means of physical activation and the use of statistical design of experiments (DOE) for activation, with the purpose of studying the effect of initial conditions on the AC production.

It is also the purpose of using these materials as adsorbents of Hg ions in water solution, describing the kinetics and diffusional aspects, by means of empiric models, in order to know the behaviour of the porous structure of AC, from the chemical and structural point of view and the nature of these processes by means of adsorption isotherms. Due to the amount of carbon available in the lignocellulosic bamboo structure, this natural resource represents a continuous source of raw material and it is an attractive material for AC production.

Experimental

DOE factorial 2^4 showed in table 1, was done in order to activate the selected carbons obtained in the previous stage. Four factors in a matrix arrange were analysed: carbon precursor (A), particle size (B), temperature (C) and activation time (D). Answer variables were the superficial area, measured as iodine number, and the SOAG content.

Table 1. DOE nomenclature for activation stage

		550 ºC		650 ºC	
		2 h	4h	2h	4h
C_{BV}	0.25 cm	CA11	CA12	CA13	CA14
	0.55 cm	CA21	CA22	CA23	CA24
C_{GA}	0.25 cm	CA31	CA23	CA33	CA34
	0.55 cm	CA41	CA42	CA43	CA44

Carbon sample with know weight, precursor carbon, particle size and indicated by DOE, was deposited in a quartz tube reactor heated by Thermolyne 21100 furnace (Di = 4.5 cm y L= 40 cm) in horizontal position. Steam water known flow was injected as activating agent and then the heating was increased in order to reach the temperature according to DOE. This temperature was kept, the time marked by DOE; and finishing this time, by natural convection the reactor was cooled to room temperature. The activated carbon obtained was washed, dried, and labeled according to DOE.

Properties of CA were measured; humidity (ASTM D 2867-99), pH (ASTM D 3838-80), density (ASTM D 2854-96), iodine number (ASTM D 4607-94), surface oxygen functional acid or basic groups by Boehm's method[6]. To analyse structural superficial features, X-ray diffraction pattern (XRPD) in the range $5 \leq 2\theta \leq 90$ has been taken on a Siemens D501 diffractometer (Cu $K\alpha_1$ radiation λ = 0.15406 nm). To analyse structural superficial features scanning electron microscopy (SEM) micrographs have been obtained with a JSM 6335 F electron microscope, operating at 10 kV and a working distance of 15 mm.

Adsorption capacity of mercury ions in water solution was made in static way under optimal conditions determined experimentally (AC dose of 0.6 g, initial concentration of Hg(II) of 10mg/L, pH of 9 and contact time of 16 h) and by means of ASTM 3500 Hg D.

Results

Table 2 shows physical-chemical properties of the AC. Seven of them present area superficial, measured as iodine number, above commercial value (550 m^2/g AC). DOE analysis my means of Daniel's method[7] had shown that the second order interaction (carbon precursor and time) modified the growth of the porosity and area superficial, obtaining the best results with the carbon provided by *B. Vulgaris Striata* and activation time of 4 h.

Table 2. Physic and chemical properties of the activated carbons

AC	Density g/mL	Humid %	pH	Yield %	Iodine number	SOAG meq/g
CA11	0.40	14.80	9.98	85.32	1500	1.03
CA12	0.35	11.00	9.87	84.14	364	1.04
CA13	0.32	31.32	9.91	74.49	502	1.25
CA14	0.33	50.96	9.95	67.39	421	0.93
CA21	0.35	9.81	9.82	85.04	590	1.01
CA22	0.33	7.71	9.81	82.34	1000	1.08
CA23	0.33	5.34	10.02	66.65	480	1.09
CA24	0.26	60.30	9.97	67.69	600	0.93
CA31	0.33	39.16	9.76	77.07	490	1.23
CA32	0.30	10.72	9.95	60.82	1340	0.86
CA33	0.29	48.71	9.97	54.26	440	0.92
CA34	0.23	67.02	9.85	20.98	1000	1.09
CA41	0.22	60.49	10.03	84.48	420	0.87
CA42	0.22	24.11	10.01	83.70	820	0.93
CA43	0.20	66.94	9.94	58.00	280	1.00
CA44	0.19	52.77	9.86	31.79	320	1.21

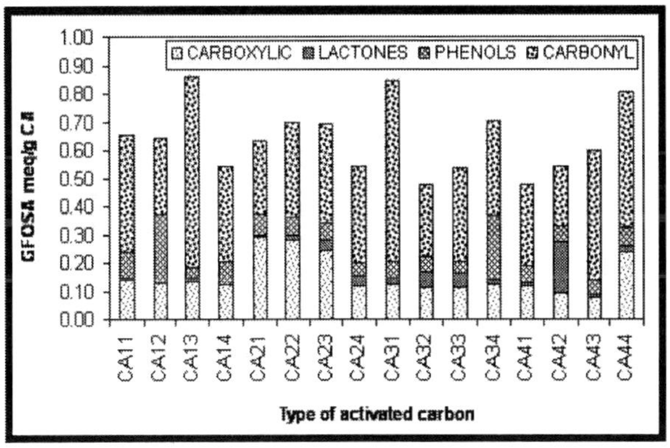

The third order interaction (carbon precursor, temperature and time) is the responsible of the FSOAG (fig. 1) development on the AC surface, resulting in higher proportion (58.57%) carbonyl groups, these kind of oxides groups modified some properties of the AC, such as pH (average of 9.91) and surface polarity; responsible of the interchange of the surface oxygen by chemical compounds with polar affinity and lightly polar. The increase of these groups was reached with *B. Vulgaris Striata* carbon, 650 ºC and 2 h.

The appearance of a broad peak centered at the 2θ angle of 28º, obtained by means of X-ray powder diffraction (fig. 2a), suggest the presence of silica[8] as remnant material after the processes of carbonization and activation, this can be due to the lignocellulosic nature of the bamboo and the relative low temperatures of thermal treatment which ones doesn't permit the complete elimination of theses siliceous materials with high boiling point. The presence of the 002 and 101 (~ 24º and 43º respectively) of the graphite reflections show the amorphous nature of the AC containing a high disordered structure with carbon atoms randomly arranged.

The resulting structure of the AC has been acquire and studied by SEM. In figure 2b, it can be observed that, even after thermal treatment the micelle shape of the cellulose present in the bamboo remains and that it is possible to find it, this kind of structure, surrounded by layers of amorphous carbon and other small pieces of fragmented layers by the long time of exposition to the activation atmosphere. It is possible to assume that in this carbon layers the resulting porosity has been developed in a perpendicular direction to the water steam flow, breaking some of the micelles and producing holes trough the walls as it can be observed in the figure 2c. Some of these holes, are not completely created due to inorganic compounds that have not been removed in the carbonization-activation stages.

From the SEM images can be observed the area superficial created in the AC, as an arrange lightly ordered among the distribution of porous size (fig. 2d). The rise in the activation time, provides the growth of the micropores until creates mesopores and eventually "small macropores" developing this ones in the treatment of 4h. These macropores, will allow the easy access of the adsorbing species inward of the macropores channels where the micropores lie.

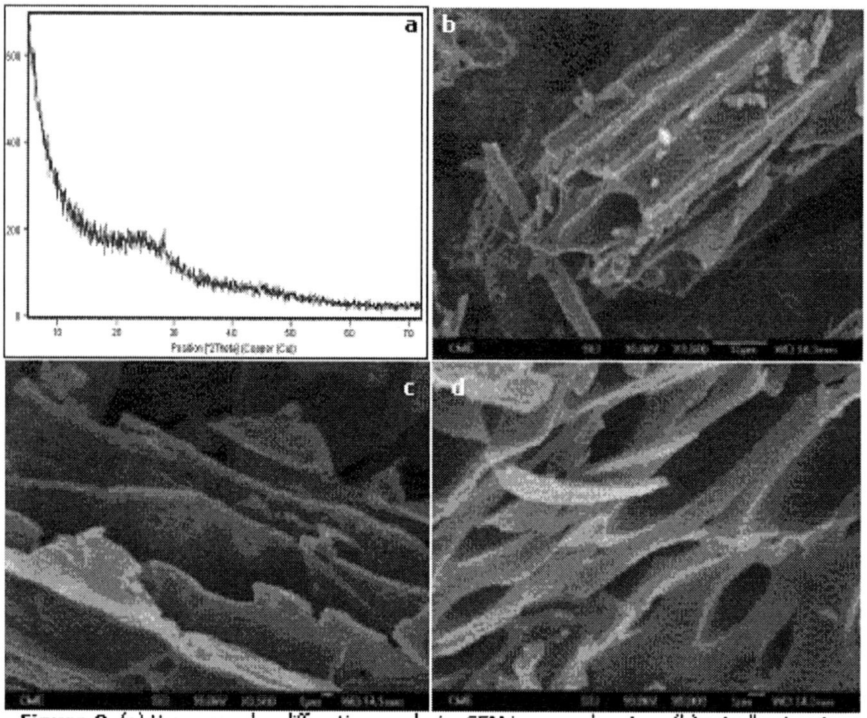

Figure 2. (a) X-ray powder diffraction analysis. SEM images showing: (b) micelle structure (c) pore structure on layered AC (d) pore size distribution.

Kinetic adsorption behaviour was adjusted to Fleming's empirical model (fig. 3a). The nature of this process follows the isotherm described by Freundlich. The shape of the isotherm (fig. 3b) shows that over the heterogeneous and amorphous surface of the material the ions do not compete by the available vacancies and suggest the formation of single layer adsorption where it can be observed the interaction between the mercury ions and the lightly polar surface of AC, the bonding energy of the ions with the FSOAG becomes in the controlling force making of this process a physical interaction between differential of electrical charges and so the diffusion process is not significant.

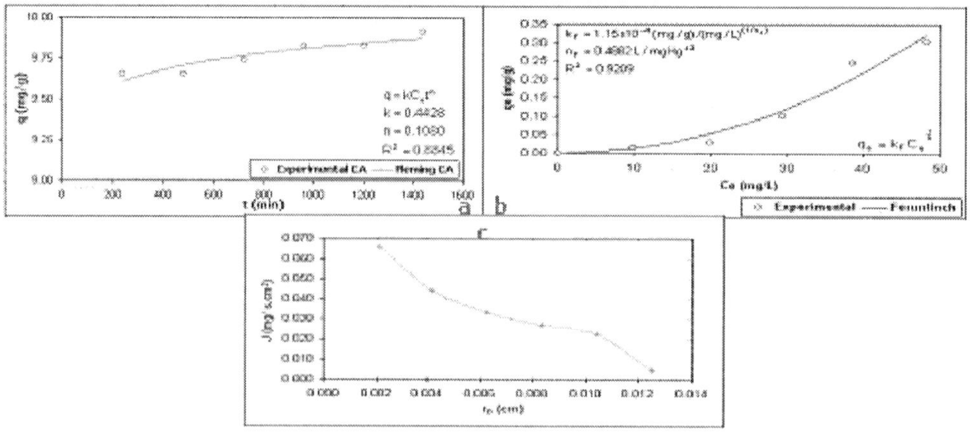

Figure 3. Experimental data adjusted to (a) empirical kinetic model (b) adsorption isotherm model (c) diffusional pore behaviour

The diffusion of the mercury ions inward of the microspores (fig. 3c) is limited by the radii of this ones, producing the filling of the internal channels. The access of the ions to the porous network can be attribute and controlled by the superficial diffusion. Due to the size and shape of the macropores network its permissible that in the free middle path of the ions the unique interaction with the AC had been physically by means of Van der Waals forces and this permit the desorption (by means of changes in the surface polarity) of the ions by chemical interchange giving the possibility of regenerate the AC and use it again in a continuous way even during five cycles more.

Conclusions

It was possible to produce AC by thermal treatment of the bamboo species *Bambusa Vulgaris Striata* and *Bambusa Oldhammi*, obtaining amorphous carbon with high area superficial and carbonyl groups content. The AC obtained shows by means of SEM the enlargement of the micropores into macropores due to the increase of the activation time, reason why the mercury process of adsorption-desorption was possible in an efficient way.

References

1. Bandosz, T.J., Bagreev, A., Abid, F. (2000). *Environ. Sci. Technol.*, 34, 1069-1074.
2. Basso. M. C., y Cukierman, A. L. (2004). *Avances en Energías Renovables y Medio Ambiente*, 8, 7-12.
3. Villegas, P. J., Rodríguez, Y. D., Wasserman, B. B. (2004). *Avances en Energías Renovables y Medio Ambiente*, 8, 53-58.
4. Boppart, S., Ingle, L., Potwora, R. J. (1996). *Chem. Proc.*, 79-85.
5. Ordóñez, V. (1999). *Madera y Bosques*, 1, 3-12.
6. Boehm, H. P. (1994). *Carbon*, 32, 759-765.
7. Montgomery, D.C. (1991) Diseño y análisis de experimentos. Capítulo 9. Grupo Ed. Iberoamericana, México.
8. Sricharoenchaikul, V. y col. (2008). *Energy & Fuels, 22,* 31–37.

337a Chemical Engineering Education Trends in Latin America

Tom R. Marrero, University of Missouri-Columbia, 1313 Fieldcrest, Columbia, MO 65203

The purpose of theis study was to assess the trends in the number of graduates from chemical engineering university programs in Latin America, including statistics about the faculties, research areas, and industrial employment. Information was obtained from the AIChE Chemical Engineering Faculty Directory(s) from 1983-present and computer data bases: Scifinder, Google, and HEGIS; when warranted specific universities were contacted directly.

The available results include data from eleven nations in 1)South America, 2)Central America, the) Caribbean, and 4) Mexico. Graduation rates from more than two dozen universities since 1983 were tabulated as a function of time for over twenty years. Preliminary results indicate that the greatest increase (two times) in chemical engineering graduation rates occurred from two universities in Brazil. Chile and Mexico each had one university that more than doubled there graduation rates over a twenty year period. This assessment describes selected programs from the four Latin American regions, above. The results of this study indicate that there is a significant opportunity for additional growth for chemical engineers in Latin America.

ARTICULATED EXPERIENCES IN PILOT PLANT

Eleonora Erdmann, Facultad de Ingeniería – CIUNSa – UNSa – INIQUI – CONICET, Salta, Argentina
Lorgio Mercado, Facultad de Ingeniería – CIUNSa – UNSa – INIQUI – CONICET, Salta, Argentina
Liliana Ale Ruiz, Facultad de Ingeniería – CIUNSa – UNSa, Salta. Argentina
Adolfo Riveros Zapata, Facultad de Ingeniería – CIUNSa – UNSa, Salta. Argentina

Abstract

Interdisciplinarity makes evident the links between the different assignments, reflecting an asserted scientific conception of the world, that shows how the phenomena does not exist in separated ways, and that when it is interrelated through the contain, an interpretation drawing is designed, interaction and dependency of the developed world. The accomplishment of an adequate relationship between the different assignments of a study plan, influences in the subsequent increase of teaching effectiveness, in quantitative as well as in qualitative terms, thus meaning a better formation of the students, which leads to a mayor qualification of the teacher. Furthermore, this constitutes a didactic condition and the challenge for the fulfillment of the scientific teaching character. The knowledge without links among them, break the conscious assimilation of knowledge and abilities. Based on the objective of the articulation between specific subjects of the Study Plan of the Chemical Engineering and to the effects of reinforcing the practical formation of the students with experiences at a pilot scale, the proposal to implement a summer course for the Chemical Engineering students with the fourth year approved is presented. These summer courses will consist in a series of Pilot Plant experiences, which are designed in such a way that the experimental results sum up information enabling an integrated project. At the same time it is intended that the student takes part in the discussion of the experience to be programmed, also when putting in operation the experimental equipment or equipments.

Introduction

We notice that in most of the cases we tend to "multidisciplinarity", which responds to the sum up organization form, the relationship between the different disciplines not clarified. See Figure 1. Interdisciplinarity appears in some cases, which are different grades of interaction between two or more disciplines. See Figure 2.

We would like to reach transdisciplinarity, which is the mayor possible grade of interaction. See Figure 3.

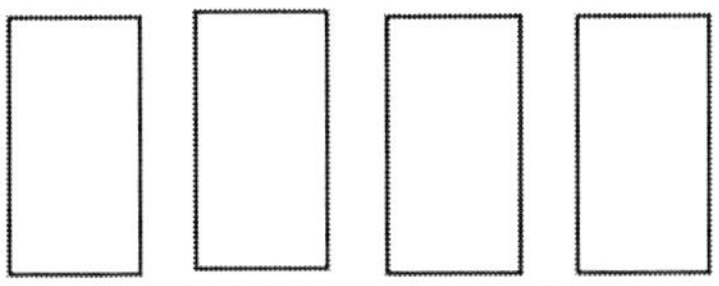

Figure 1. Multidisciplinarity Scheme

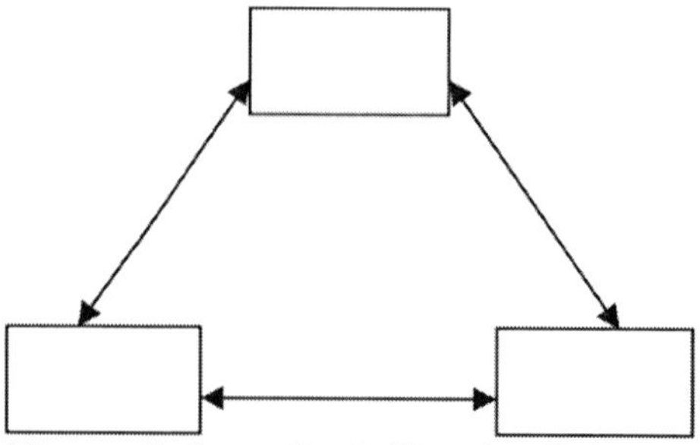

Figure 2. Interdisciplinarity Scheme

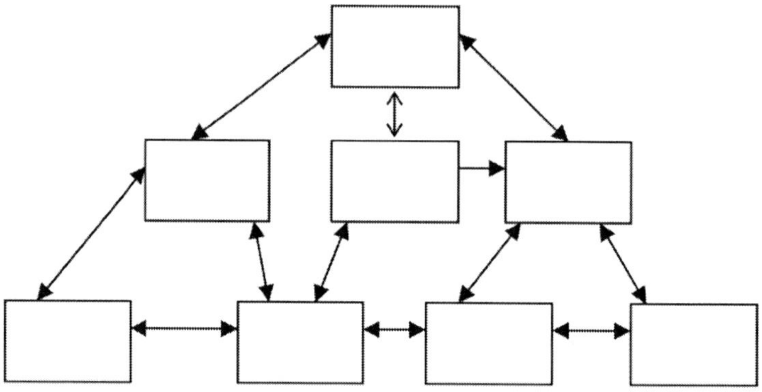

Figure 3. Transdisciplinarity Scheme

The Articulation

According to Tamayo and Tamayo (2004) interdisciplinarity is defined as the conjunction of linked disciplines among themselves with defined relationships, with the purpose that its activities do no happen in an isolated, disperse and fractioned way.

The evidence of links among the different assignments of the career which reflect an asserted scientific conception of the world, showing how the phenomena do not exist in separate ways and that by interrelationing through contains, and a draft of interpellation, interaction and dependency of the world's development is designed.

In order to reach an adequate relationship between the different study plan assignments influences the subsequent increase of effectiveness of the teaching in quantitative terms as well as in qualitative terms. That means an optimal preparation of the students, which requires a mayor qualification of the teacher. Knowledge without links among themselves, break the conscious assimilation of knowledge and abilities. Furthermore, this represents a didactic condition and the requirement for the accomplishment of the scientific character of teaching.

Krupskaia pointed out about this matter, *"if we want to transmit to the teaching staff the essence of the scientific conception of the world, we should not teach incoherent knowledge, but with its corresponding and determined relationship among them, only in this sense, we conceive a more integral character than this one".*

We understand under articulation the relationship between various assignments of different disciplines, without loosing the logical structure required for ordering the contents, warranting the development of specific abilities of the assignments from the professional and those that make up the personality of the student.

Articulation among the assignments should start from the general starting point of the interrelation or concatenation of all phenomena which are reflected in teaching through the contents and themes that are common to various assignments.

Theory of the Significant Apprenticeship

Ausubel states that the apprenticeship of the student depends on the previous cognitive structure related with the new information, "new structure", should be understood as the conjunction of concepts, ideas which the individual person owns in a given area of knowledge, as well as its organization.

Ausubel resumes this fact in the epigraph of his book, as follows: "If I had to reduce the whole educational psychology into one single first step, I would enounce this one: The most important factor which influences the apprenticeship is that, what the student already knows. Find it out and teach consequently" (AUSUBEL; 1983).

Significant and Mechanical Apprenticeship

The most significant characteristic of apprenticeship is the fact that it produces interaction between the most relevant knowledge of the cognitive structure and new information (it is not a simple association), in such a way that these acquire a significance and are integrated to the cognitive structure in a non-arbitrary and substantial way, for the benefit of differentiation, evolution and stability of the pre-existing subsunsors (each time new information "is connected" with a relevant concept) and subsequently of the whole cognitive structure.

Contrary to the significant apprenticeship, the mechanical apprenticeship, is produced when adequate subsunsors do not exist in such a way that the new information is not properly saved, without interacting with existing knowledge, an example of which would be the simple learning of formulas in physics, this new information is incorporated to the cognitive structure in a literal, not proper way because it consists of purely arbitrary associations, [when], "the student shows lack of relevant and necessary knowledge to enable that the learning task might be potentially significant" (independent of potential significance that the task might have)... (Ausubel; 1983).

By way of illustration, in physics, if the system concepts of work, pressure, temperature and energy conservation do already exist in the cognitive structure of the student, these will serve as subsunsors for new knowledge, about thermodynamics, such as thermal machines, steam turbines, fusion reactors or simply the basic theory of the refrigerators; the process of new interaction with the already existing one produces a new modification of the subsunsors concept (work, energy conservation, etc.), this implies that subsunsors are wide, clear, stable or unstable concepts. All these depend on the way and frequency they are exposed to interaction with new information.

In the example given, the idea of energy conservation and mechanical work, will serve as "anchorage" for new information about thermal machines, but as soon as these new concepts are learnt in a significant way, they will grow and will modify the initial subsunsors, that is to say, the energy conservation and mechanical work concepts would develop to serve as subsunsors for concepts such as the second thermodynamic law and entropy.

Finally Ausubel does not state a distinction between significant and mechanical apprenticeship as a dichotomy, but as a "continuum", and more, both apprenticeship types can incur concomitantly in the same learning task (Ausubel; 1983); for example, the simple memorization of formulas would situate in one end of this continuous (mechanical apprenticeship) and the learning of relations between concepts could situate on the other end (significant apprenticeship).

In brief, the significant apprenticeship is to develop abilities, allowing the student to work out knowledge in a concrete, functional form, affecting his profession's duties accomplishing more productive and creative kinds of work. Furthermore, it means to gain that each developed action leads to a higher degree of skill of the contents, by means of transforming primary characteristics of said actions.

Theory Design

Problem

For Chemical Engineering students of 5th grade, at the National University of Salta`s Engineering Faculty. Is it possible to accomplish a significant apprenticeship upon the basis of Ausubel's theory, by means of a change in the teaching methodology, which considers the links in the study assignment plan?

General Purpose

To institute a new teaching methodology, that contributes to improve the apprenticeship of the students.

Specific Purpose

To design and put in practice the experiences of the pilot plant which should consider the assignment links of the study plan.

Purpose of the Study

Teaching procedure – apprenticeship.

Action Field

- Significative Apprenticeship.
- Articulation among assignments.

According to the theoretical foundation, the analysis of the Object and the Objects of the investigation, we state following hypothesis.

Hypothesis

The application of practical Works which considers the links of the assignments in the study plan will contribute in the accomplishment of significant apprenticeship at the pilot plant.

Practical Articulation of Specific Subjects in Chemical Engineering

Based on the articulation purpose among specific subjects of the study plan in the Chemical Engineering Career and to the effects to reinforce practical formation of the students by means of multidisciplinary experiences, designed upon procedures at pilot scale, the proposal is presented to institute a summer course directed to the Chemical Engineer students, who have approved the 4th year.

During the institution of this summer course the student is expected to take part in the discussion of operational alternatives of the programmed procedure, putting in operation the experimental equipment and the analysis of obtained results.

This proposal is designed with the purpose of articulating subjects of the curricula, such as Organic Chemistry, Kinetic Chemistry, Physical Chemistry, Operations and Procedures, Auxiliary Services, Instrumentation and Control, Clean Production, etc., in such a way that the students use the knowledge granted and acquired during the whole career, with the final purpose of contributing to integrate concepts and practical formation.

During the course of the several assignments, the students carry out routine labor and pilot plant tests on specific themes of each one, knowing in separate ways the functioning and operation of most of the available equipment in the pilot plant. During the summer course the function of the plant will be integrated on a total basis, the students will have to apply their operational and functional knowledge of each one of the equipments and the procedure problematic as a whole.

Actions to follow

- Request the Chemical Engineering Career lectures for execution alternatives of new practical labor and/or pilot plant tasks.
- Analysis of the proposal of practical and/or pilot plant labor with the objective of acquiring data about:
    - Indispensable reactives
    - Indispensable equipment
    - Instrumental required
- Analysis of the feasibility of the experiences in an integrated way, within an admissible period of time.
- Adjustment and interrelation of the selected labor and pilot plant tests.

Available Equipment

In order of accomplishing lecture interrelation, the Engineering Faculty has the Pilot Plant II available, with the specific purpose of developing academic activities.

Several equipments is available, most of which at pilot scale see (Figure 4), and periphery laboratories are also available.

Following is the equipment available:

- Water softening system, 100 l/h.
- Steam tubular boiler, 100 kg/h, 7 kg/cm2
- Refrigeration system with 8 m3 storage depot, 200 l cooler hole.
- Cooling tower, forced fan, 15 TR.
- Stirred tank, 150 l, and ½ HP stirrers.

- Ultra high temperature sterilization system.
- 140 l heater for water tank.
- Liquid-gas absorption equipment.
- Liquid-liquid extraction system.
- Load loss rack.
- Air compressor.
- Jaw crusher, Blake model.
- Cone crusher, Symond model.
- Hammer mill.
- Mesh sieve.
- Rotative oven for calcination.

Laboratories of electrical installations, Automatic control, Electronic, Petro chemistry, Operations and Procedures, Clean production and Computer science teaching for procedure simulation.

Furthermore there is availability of a series of tools, working table and personal protection elements.

Figure 4 – Pilot Plant II equipment: Softener, Cooling tower, UHT, Absorption tower, Crusher

Articulation Examples

Obtention of precipitated Calcium Carbonate (PCC) from exhausted brine during the water softening phase.

The practice involves the interplay of four assignments for the development of labor, beginning with the operation of water softening using exchange resin. Once the resin is exhausted, the resin is regenerated of it with brine (12% in ClNa). A exhausted brine is unloading As a result of this operation (rich in Ca and Mg) to be fitted properly and then be ready

The proposal is to treat the effluent before its discharge into the environment, in order to recover the brine for reuse in the regeneration process of exchange resin and also to obtain viable by products marketing

The depleted brine is treated in a batch stirred tank reactor with appropriate reagents, such as Solvay Soda and Caustic Soda, so remove impurities differentially (Ca and Mg).

The suspension is then filtered, and each solid is drying properly

Involved chairs and objectives

The involved chairs in the development of practical and theoretical concepts to integrate them are the following:

- Auxiliary Services and Unit Operations I: Water Treatment (softening), pumps, agitation, filtration.
- Clean Production: treatment and disposal of effluents
- Operations and Processes: reactors design and design of dryers.
- Instrumentation and Process Control: design of instrumentation and control system of each of the integrated phases

Diagrams of the pilot plant practices

A diagram of the conventional process in water softening plants, by cation exchange resins, sodium cycle is presented in Figure 5.

In Figure 6 represents schematically the alternative process to develop in practice, with recovery of brine and obtaining products of commercial application

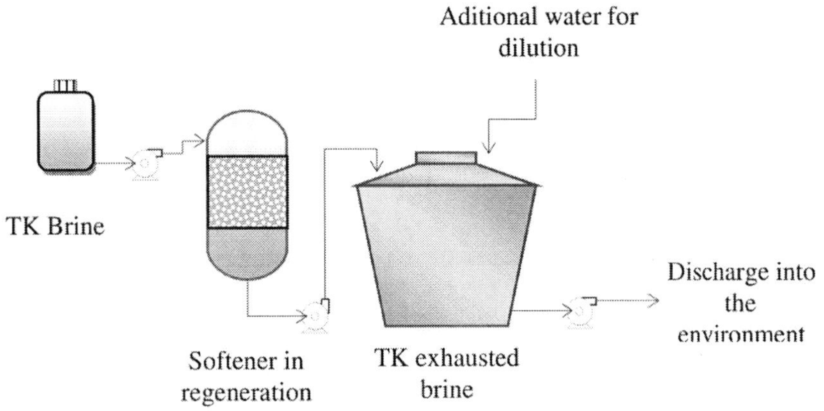

Figure 5. Conventional scheme for the regeneration operation of a cation exchange resins, sodium cycle

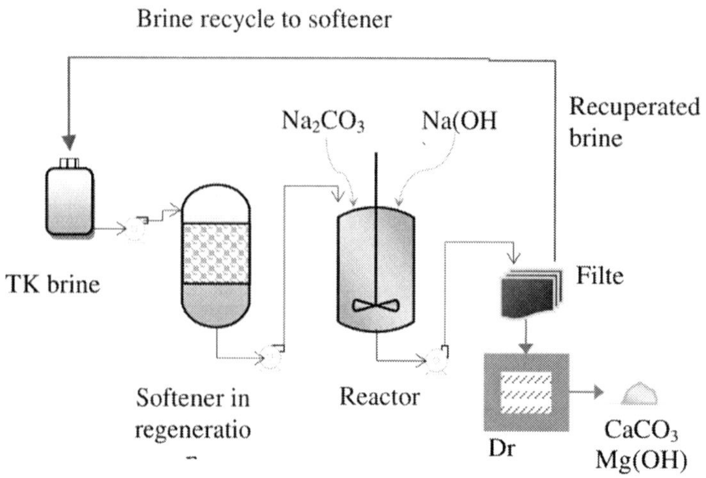

Figure 6. Scheme designed for the integrative practice, with recovery of brine and obtaining by products of commercial application

Saponification of ethyl acetate with caustic soda

This practice is to design a batch reactor to achieve a conversion determined from kinetic data of an irreversible and in non-homogeneous catalytic phase reaction.

The reaction of saponification is as follows:

$$CH_3COOCH_2CH_3 + NaOH \longrightarrow CH_3COONa + CH_3CH_2OH$$

Involved chairs and objectives

- Organic chemistry: a study of the reaction of saponification.
- Chemical Kinetics: determination of the kinetics of the reaction (laboratory scale)
- Operations and Processes: Design of a batch reactor to achieve a determined conversion from the kinetic data.
- Instrumentation and Control: defining the instrumentation for the measurement and control of operational variables.

The outline of the experience in the pilot plant can be observed in Figure 7.

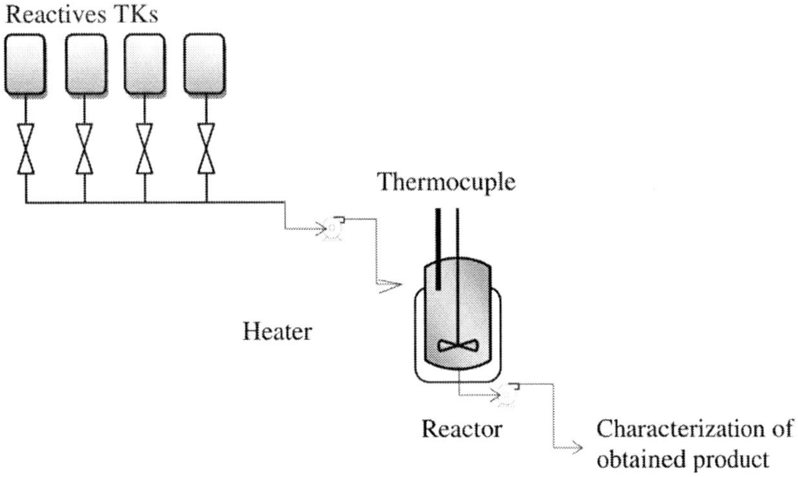

Figure 7. Outline of Practice for the saponification of ethyl acetate

Conclusions

For all these reasons, we believe that the realization of practical integrated in pilot plant for the career of Chemical Engineering, Faculty of Engineering will allow:

- Articulate specific areas of the career.
- Strengthening the basic knowledge by conducting a pilot scale integrated practices.
- Give more emphasis on practical training for students in the last year of the career.
- Increasing the participation of pupils in the programming of practical experience in order to generate critical discussions in the construction equipment and the analysis of results

References

1. Ausubel,Novak & Hanesian (1983), "Psicología Educativa: Un punto de vista cognoscitivo", 2° Ed. Trillas México,
2. BetzDearbon, (1991), "Handbook of Industrial Water Conditioning", 9th Ed.
3. Brown, G.G. (1956), "Operaciones básicas de la Ingeniería Química", Manuel Marin Ed. Barcelona (España).
4. Carberry J., (1980), "Ingeniería de las reacciones químicas y catalíticas", Mc Graw Hill.
5. Cunningham R. & J., Lombardi, (1972), "Fundamentos del diseño de reactores", EUDEBA.
6. Debingh, K. G., et al, (1990), Introducción a la teoría de los reactores químicos; Limusa Noriega.
7. Durst, H. & G., Gokel, (1985), "Química Orgánica Experimental", Editorial Reverté.
8. Froment & Bischoff, (1979), "Chemical Reactor Analisysis & Desing", John Wiley & Sons.
9. Hardwood, L., (1999), "Experimental Organic Chemistry: Standard and Microescale", 2º Ed., Blackwell Science.
10. Henley, E. et al, (1988), "Operaciones de separación por etapas de equilibrio en Ingeniería Química"; Reverté
11. King C., (1981), "Procesos de separación", Reverté.
12. Levenspiel, O, (1974), "Ingeniería de las reacciones químicas", Reverté.

13. Mc Cabe, W. & J., Smith, (1967), "Unitaries Operations of Chemical Engineering". 2da.Ed., Mc. Graw-Hill Book Co, USA
14. Novak, J & B, Gowin, (1988), "Aprendiendo a Aprender", Martínez Roca.Barcelona
15. Rase, H. , (1977), "Chemical Reactor Desing for Process Plants", Vol I y II, John Wiley & Sons.
16. Richardson, J. & D.G. Peacock, (1971), "Chemical Engineering", Primera Edición.
17. Scott Fogler, H., (2001), "Elementos de Ingeniería de las Reacciones Químicas", 3° Ed. Pearson Educación, México.
18. Smith, J., (1981), "Ingeniería de la Cinética Química", Continental, S.A., México, Segunda Edición
19. Tamayo y Tamayo, M., (2004), "Diccionario de Investigación Científica", Editorial Limusa, México, Segunda Edición.
20. Vian Ortuño, A., (1996), "Introducción a la Química Industrial", Reverte.
21. Wade, L., (1995), "Química Orgánica", Prentice Hall.

337c Nanobiotechnology Education and Outreach at the College of Nanoscale Science and Engineering: A Molecular Frontier in Chemical Engineering

Yubing Xie, Nathaniel C. Cady, and Diana Martin. College of Nanoscale Science and Engineering, University of Albany, 255 Fuller Road, Albany, NY 12203

Traditional chemical engineering is the branch of engineering for the development and application of processes that change materials based on chemistry, physics and mathematics. Chemical engineering continues to evolve with biological sciences and nanotechnology. Emerging nanobiotechnology brings a molecular frontier for chemical engineering. Education and outreach activities in nanobiotechnology at the College of Nanoscale Science and Engineering of University at Albany will be introduced. A few examples to expose K-12 and undergraduate students to nanobiotechnology will be presented. Chemical engineering principles in these engineered nanobiological systems will be identified. A "top-down" modular approach to train graduate students, undergraduate students and high school students and involve K-12 students in science and engineering will be discussed.

New Tools for Teaching Chemical Engineering Thermodynamics

J. Richard Elliott
Chemical and Biomolecular Engineering Dept.
The University of Akron
Akron, OH 44325-3906
jelliott@uakron.edu

Abstract. The teaching toolbox described by Elliott and Lira (2000) has been expanded to include three new tools: molecular simulation, ConcepTesting, and a simplified version of the MOSCED model. The previous list of tools included: detailed derivations (e.g. Maxwell's relations), computational tools (calculator and xls), projects, homework, analogies, examples, tours, tests (including samples from past years), quizzes, and help sessions. Students were surveyed to rank these tools from "most instructive" to "least." The new tools are described briefly and the survey assessments are presented. The molecular simulation tool focuses on applets posted at: http://rheneas.eng.buffalo.edu/wiki/DMD. These applets provide visualization of molecular dynamics for ideal gases, hard spheres, and square-well spheres. The students are guided through several homework assignments in which they learn about temperature, energy, pressure, and system size. Further details are available online, so the remainder of this abstract focuses on the Simplified Separation of Cohesive Energy Density (SSCED). In the current work, the acidity and basicity parameters are adopted directly from the latest literature, but the polarity and dispersion parameters are lumped together and the total of all contributions is constrained to match the original Scatchard-Hildebrand solubility parameter. Three composition-dependent parameters of the MOSCED model are set to constants. In this way, the contrast between the physical interactions and the chemical interactions is more readily apparent and the model can be applied directly at all compositions in a self-consistent manner. Examples are given of in-class and exam assessments, along with homework assignments. The simplified model is intended to make the key concepts of hydrogen bonding accessible to any college student, including freshmen. ConcepTesting is demonstrated with in-class assessments of the SSCED model. ConcepTesting refers to an interactive form of class engagement especially as it applies to the use of student response "clickers." Students are prompted with multiple choice questions and their responses are instantly collected and displayed electronically. Students are encouraged to work in small groups (~3) in developing their responses. A strict interpretation of ConcepTesting would limit questions to abstract conceptual content. In our implementation, we integrate conceptual content with more conventional examples and problem solving. This leads to a class that is continuously engaged.

Keywords: Activity coefficient, ConcepTesting, MOSCED, molecular dynamics.

1. Introduction

It seems that the freshmen are onto me. As part of our freshman "Chemical Engineering Computations" course, the students can choose speakers for four half-lectures. Last Spring, they chose me to give a presentation about hydrogen bonding. Rumor has it that they heard of a fanatic professor who was particularly fanatic about thermodynamics, with a soft spot for the topic of hydrogen bonding. Their strategy was two-fold: first, to butter this guy up by asking him to ramble on about his favorite subject and second, to scout the demon of this dangerous domain called thermodynamics. In my turn, I saw this as a teachable moment. If I made the presentation sufficiently accessible, they might accidentally learn something about thermodynamics.

But what computational model can be accessible to freshmen in 25 minutes and explain hydrogen bonding and its role in chemical engineering? A little scouting of my own revealed that they were already performing flash computations. So they knew about $K_i = y_i/x_i$, but the only solution model they had was Raoult's Law ($K_i = P_i^{sat}/P$). This was my way in. According to Raoult's law, $y_E > x_E$ for ethanol in water at all compositions. Then distillation to gasohol should be no problem because it is constantly enriching. But experiments show that $y_E < x_E$ when $x_E > 0.9$. If you put 10% water into your gas tank, water separates and you have a major problem on your hand.

The reason this happens is that the water does not entirely "like" the ethanol. We can characterize this disdain with modified Raoult's Law ($K_i = \gamma_i P_i^{sat}/P$). If a little ethanol mixes with the water, that is fine and γ_W=1. But when water is surrounded by 90% ethanol, it can get very uncomfortable and $\gamma_W \rightarrow 2$. At $x_E = 0.9$ and 78°C, ethanol is comfortable and $K_E = P_E^{sat}/P \approx 1$ but $K_W = \gamma_W P_W^{sat}/P \approx 1.1$. So the water becomes richer in the vapor than the ethanol and distillation fails. This causes the separation to make gasohol to be less direct and more expensive both in terms of dollars and in terms of energy efficiency. They nod when I say this, but five minutes have already elapsed.

In the remaining 20 minutes, I introduce a simplified version of the MOSCED model and apply it in an example. Then I assign two homework problems. I refer to this model as the SSCED (simplified separation of cohesive energy density) model. The MOSCED model is a modification of Scatchard-Hildebrand theory that separates the cohesive energy density into a dispersion term, a polarity term, and two hydrogen bonding terms (one for acidity and one for basicity).[1] The simplifications of SSCED are designed to convey key concepts in a manner that is consistent with presentations throughout thermodynamics. Quantitative precision is not necessary for this conceptual introduction. On the other hand, the interpretation of a term like the binary interaction parameter, k_{ij}, is fundamental and intellectually challenging to students. The concept that $k_{ij} < 0$ indicates favorable mixing can be illustrated graphically with the square-well potential, and reinforced with ConcepTesting[2] as demonstrated in the assessment section below. The SSCED model reinforces that concept while the Scatchard-Hildebrand model does not, as discussed in the rationale section along with other finer points about the advantages and limitations of the SSCED model. The presentation section immediately after this introduction closely follows the notes for the freshman lecture. I conclude with a brief review of the assessments of student learning and a perspective on how students may benefit from presentation of such a model at an early stage in the curriculum.

2. Presentation

Resuming from the first five minutes of the lecture, the technical name for the factor, γ_i, is the "activity coefficient." When $\gamma_i = 1$ the situation of the ith component is "ideal." When $\gamma_i < 1$, the ith component is extremely comfortable. Such a formulation would make a great solvent if you had a nasty stain to remove. When $\gamma_i > 1$, the component is uncomfortable, like the water in 90% ethanol. Finally, when $\gamma_i > 10$, the component "hates" its environment so much that it separates, like the water in gasoline. A very simple computational model can describe all of these situations and help to design formulations to achieve chemical engineering goals. It is,

$$RT\ln\gamma_1 = V_1(1-\Phi_1)^2[(\delta_2' - \delta_1')^2 + 2k_{12}\delta_1'\delta_2'] \qquad (1)$$

Where R = 8.314 J/mole-K, T is the temperature in Kelvins, $V_i = MW_i/\rho_i^L$ is the liquid molar volume at 298K and $\Phi_i = x_iV_i/\sum x_jV_j$ is the volume fraction, analogous to weight fraction. The bracketed terms require some explanation. k_{12} is a correction factor that characterizes specific interactions, principally hydrogen bonding. We discuss k_{12} later. The other term addresses the "solubility parameter." If we assume for the moment that $k_{12} = 0$, then $\Delta\delta'$ provides a concise and quantitative measure of γ_i. If $\Delta\delta' = 0$, then the solution is ideal and Raoult's Law is fine. Otherwise, the solution becomes non-ideal.

The solubility parameter is related to the energy density of a compound. This energy can be quantified by the heat of boiling. When you boil water, for example, molecules are extracted from their congenial environment to a lonely vapor, where they can share no energy with others. They prefer to share energy. That is why you must add heat. More heat must be added if they share more energy. If the same boiling pot is used to characterize various compounds, then more small molecules will fit in it than large ones, and even more heat is required. Therefore, it is the energy density that characterizes how strongly a compound sticks to itself. This kind of energy density is something quite different from the explosive energy density of a compound like dynamite, so we need a distinctive name for it. That name is the cohesive energy density, defined by,

$$\delta^2 = (U^{vap}/V) = \text{J/cm}^3 \qquad (2)$$

where U^{vap} is the energy of vaporization and the rationale for squaring δ is explained in the thermodynamics course. In the absence of hydrogen bonding, $\delta = \delta'$. Therefore, in terms of δ', discomfort of a component in solution is not caused by dislike for the other components, but by a strong preference for its own company. You may have heard that an extroverted engineer is one who looks at *your* feet when he is talking to you. Any introverted engineers in the room should relate to this perspective on the definition of discomfort. It is a sad perspective, but true.

In the presence of hydrogen bonding, the hydrogen bonding contribution must be separated from δ, hence the name for this model as the SSCED model (simplified separation of cohesive energy density model, pronounced "sked" like sled with a "k"). This separation is given by

$$\delta^2 = (\delta')^2 + 2\,\alpha\beta \qquad (3)$$

Where α characterizes the compound's acidity and β characterizes the basicity. The acidity and basicity can be measured spectroscopically by probing how strongly compounds interact with a standard reference base and a standard acid. Sample values of δ', α, and β are given in Table 1. Note that water stands out in Table 1 as a compound with remarkably high energy density, both in terms of δ' and in terms of $\alpha\beta$. The water molecule is very small, essentially the size of a single oxygen atom, but it has a large dipole moment (reflected in δ') and strong hydrogen bonding (reflected in $\alpha\beta$).

We now return to the quantity k_{12}. If k_{12} =0, then $\gamma_i > 1$, always, but there are situations when $\gamma_i < 1$. When mixing acids with water for example, the compounds "like each other" so much that you need to be careful. A more moderate example is given by mixing acetone with chloroform, in which case mixture boiling experiments show that $\gamma_i < 1$. The proton of the chloroform is made mildly acidic by the electronegative chlorine atoms pulling on its electrons. The high density of electrons on the carbonyl oxygen makes it mildly basic. Organic chemistry courses should reinforce these concepts of electron distributions. These considerations are represented by the guideline that

$$k_{12} \approx (\alpha_2-\alpha_1)(\beta_2-\beta_1)/(4\delta_1\delta_2) \quad (4)$$

Table 1. Sample values of physical properties.

	T_c(K)	P_c(MPa)	ω	C_P/R	*MW*	ρ_{298}	δ(J/cm³)$^{1/2}$	α(J/ cm³)$^{1/2}$	β(J/ cm³)$^{1/2}$
Benzene	562.2	4.90	0.211	9.82	78	0.87	18.65	0.63	2.24
Methanol	512.6	8.10	0.566	5.28	32	0.79	19.25	17.43	14.49
Ethanol	516.4	6.38	0.637	7.88	46	0.79	18.67	12.58	13.29
Acetone	508.2	4.70	0.306	8.96	58	0.79	19.64	0.00	11.14
Chloroform	536.4	5.40	0.216	7.90	119.2	1.48	18.88	5.80	0.12
Water	647.3	22.12	0.344	4.04	18	1.00	27.94	50.13	15.06
Iso-octane	544.0	2.57	0.303	22.50	114	0.70	14.11	0	0
MTBE	497.1	3.43	0.266	15.37	88	0.74	15.17	0	7.4

For the chloroform+acetone example, this formula gives

$$k_{12} = (5.8\text{-}0)(0.12\text{-}11.14)/(4*19.64*18.88) = -0.035 \quad (5)$$

Note how the order of subtraction results in a negative value for k_{12} when one of the components is acidic and the other is basic. If you switched the subscript assignments, then $\Delta\alpha$ would be negative and $\Delta\beta$ would be positive, but k_{12} would still be negative. This negative value makes the value of γ_i smaller, and that is basically what happens when hydrogen bonding is favorable.

Something else happens when one compound forms hydrogen bonds but the other is inert. Taking iso-octane(1) as representative of gasoline and mixing it with water(2),

$$k_{12} = (0\text{-}50.13)(0\text{ -}15.06)/(4*27.94*14.11) = 0.479 \quad (6)$$

This large positive value will add to the large $(\Delta\delta')^2$ such that $\gamma_1 >>10$, indicating the liquid phase split that we anticipated. We can quantify the phase split by noting that $x_i \approx 1/\gamma_i$ when $\gamma_i > 100$. Knowing the saturation limit of water contaminants can be useful in environmental applications.

As a final example, note that we recover an ideal solution when both components hydrogen bond, as in the case of methanol+ethanol.

$$k_{12} = (17.43\text{-}12.58)(14.49\text{-}13.29)/(4*19.25*18.67) = 0.003 \quad (7)$$

In this case, we see that hydrogen bonding itself is not the cause of solution non-ideality. A mismatch of hydrogen bonding causes non-ideality.

Applications of these insights abound in chemical engineering. For example, what third compound could you add to ethanol+water to make the solution more ideal so that pure ethanol could be obtained? The extension of the SCED model to multicomponent systems is simple, as discussed in the thermodynamics course. How soluble is vitamin-C in the blood stream relative to its solubility in body fat? What about vitamin-E? Aspirin? Tylenol? You just need to know the activity coefficients of these compounds in water and n-octanol (a reasonable approximation of body fat). What solvent should you use to safely remove an undesirable embellishment from a classical painting? The embellishment probably used a different paint, so you need to find a solvent with $\gamma_i < 1$ in the embellished paint but $\gamma_i > 1$ in the classical paint. Quantitative understanding of fields from art restoration to zoology to agribusiness would be impossible without unifying concepts like hydrogen bonding. Students should retain these concepts and reinforce them as they take complementary courses throughout their curriculum.

Example 1.

Estimate the K-value for 10mol%chloroform in 90% acetone at 350K and 0.1MPa. You may assume that $\log_{10}(P^{sat}/P_c)=7(1+\omega)(1-T_c/T)/3$.

Solution:
The value of k_{12} = -0.035 is given by Eq. (5). V_C = 119.2/1.48= 80.5 and V_A = 58/0.79 =73.4.
The volume fraction is: Φ_C = 0.1*80.5/(0.1*80.5+0.9*73.4) = 0.109
γ_C = exp{ 80.5*(1-0.109)²*[(19.64-18.88)² - 2*0.035*18.88*19.64]/(8.314*350) } = 0.573.
P_C^{sat} = 5.40*10^(7*1.216*(1-536.4/350)/3) = 0.166 MPa.
$K_C = \gamma_C * P_C^{sat}/P$ = 0.573*0.166/0.1 = 0.951

Homework 1.

Gasohol is made by distilling a solution known as beer (~5mol% ethanol). Compute the K-values of ethanol and water at 5mol% ethanol and 365K and compare them to the K-values at 95mol%ethanol and 350.6K. Assume that P = 0.1MPa. Explain the impact of activity coefficient on your results. You may assume that $\log_{10}(P^{sat}/P_c)=7(1+\omega)(1-T_c/T)/3$.
Solution: k_{12} = (12.58-50.13)(13.29-15.06)/(4*18.67*27.94) = 0.032
At 358.5K and 5% ethanol, P_E^{sat} = 0.1327 and P_W^{sat} = 0.0658 MPa according to the assumed vapor pressure equation. The water is nearly pure and computation confirms that γ_W = 1.016.
Details for ethanol: Φ_E = 0.05*58.5/(0.05*58.5+0.95*18) = 0.146.
γ_E = exp{ 58.5*(1-0.146)² [(18.67-27.94)² + 2*0.032*18.67*27.94]/(8.314*358.5) } = 5.517.
This gives K_E = 7.319 and K_W = 0.669.
At 350.6K and 95% ethanol, P_E^{sat} = 0.0997 and P_W^{sat} = 0.0491 MPa according to the assumed vapor pressure equation. The ethanol is nearly pure and computation confirms that γ_E = 1.001.
For water,
γ_W = exp{ 18*[(18.67-27.94)² + 2*0.032*18.67*27.94]/(8.314*350.6) } = 2.046
This gives K_E = 0.998 and K_W= 1.002.
Overall, non-ideality makes distillation easier at low concentrations of ethanol, but the large activity coefficient switches to the water at high ethanol concentrations and makes it slightly more volatile. This is why distillation fails to purify ethanol this system.

Homework 2.

The American experience with methyl tert-butyl ether (MTBE) in the 90's broaches qualification as a fiasco. Rumor has it that congressmen from corn states thought that a mandate for 10% "oxygenated fuel" would boost demand for ethanol. Within four years MTBE was the number one synthetic chemical produced in the world. What nobody anticipated was how MTBE might affect groundwater. It imparts a bitter taste and nasty smell even at parts per billion. Gasoline is stored in underground tanks, and the tanks leak. Estimate the solubility of MTBE in water at 298K and compare it to that of iso-octane and benzene.
Solution: For iso-octane, Eq. (6) gives k_{12} = 0.479.
$x_O = 1/\gamma_O$ = 1/exp{ 114/0.70*[(14.11-27.94)² + 2*0.479*14.11*27.94]/(8.314*298) } = 5.7E-17
For benzene, k_{12} = (0.63-50.13)(2.24 -15.06)/(4*27.94*18.65) = 0.304
x_B=1/exp{ 78/0.87*[(18.65-27.94)² + 2*0.304*18.65*27.94]/(8.314*298) } = 4.0E-8
For MTBE, k_{12} = (0-50.13)(7.4 -15.06)/(4*27.94*15.17) = 0.226
x_M=1/exp{ 88/0.74*[(15.17-27.94)² + 2*0.226*15.17*27.94]/(8.314*298) } = 1.7E-9

So the solubility of MTBE is roughly 8 orders of magnitude higher than that of iso-octane. This estimate would need to be checked with experimental data, but the essential observation is that the basicity of the ether suggests checking it out. Benzene is interesting because its estimated solubility is even higher than that of MTBE. Benzene does not taste or smell like MTBE, but it is carcinogenic. Nevertheless, nobody seems to be talking about the solubility of benzene in groundwater… at the moment.

3. Rationale

The SSCED model provides a simplified and generalized introduction to the MOSCED model, but the MOSCED model is designed for other purposes. Specifically, it is designed for infinite dilution activity coefficients instead of being a solution model at all concentrations. The MOSCED model is given by,

$$\ln\gamma_2^{\infty} = \frac{V_2}{RT}\left[(\lambda_2-\lambda_1)^2 + \frac{q_1^2 q_2^2(\tau_2-\tau_1)^2}{\psi_1} + \frac{(\alpha_2-\alpha_1)(\beta_2-\beta_1)}{\xi_1}\right] + aa\ln\left(\frac{V_2}{V_1}\right) + 1 - \left(\frac{V_2}{V_1}\right)^{aa} \quad (8)$$

Where λ_i is the dispersion factor, τ_i is the polarity factor, q_i is a factor ranging from 0.9 to 1. aa, ψ_i and ξ_i are adjustable parameters characterizing solvent properties. At infinite dilution, they are specific values, but they must depend on composition to change from one solvent to the next. That composition dependence is not addressed by the MOSCED model, but it poses no problem for experts in thermodynamics. Parameters of an activity model like UNIQUAC could be determined from the infinite dilution activity coefficients and activity coefficients at all compositions computed from UNIQUAC. But activity models like UNIQUAC tend to be covered after models like MOSCED. The MOSCED model is based primarily on van der Waals mixing. The terms involving $(\Delta\lambda)^2$ and (V_2/V_1) comprise the Scatchard-Hildebrand and Flory-Huggins contributions derived from the van der Waals equation when constant packing fraction is assumed.[3] The other contributions are based on phenomenological arguments. The UNIQUAC model is based on the concept of local compositions. Developing the nuances of fitting parameters to one activity model then interpolating the free energy based on an entirely different activity model could undermine the attention span of sophomores as well as freshmen. Another alternative would be to articulate composition dependencies for all the parameters. This would take the model away from a simple explanation. On the other hand, eliminating the polarity factor (ie. $\psi = \infty$), setting $aa = 0$, and $\xi = 2$ for all compositions makes the model much simpler and more broadly applicable while retaining the separation that enables consideration of hydrogen bonding influences.

The SSCED model was derived from the MOSCED model by minimizing deviations in a somewhat crude manner. The factor of 2 multiplying $\alpha\beta$ in Eq. (1) was determined by minimizing the differences between the physical contributions of the two models. Analyzing the physical contributions showed that $\delta^2 \approx (\lambda^2 + \tau^2/2)$ for non-associating compounds. Note that $\delta = \delta'$ in the absence of association. For associating compounds, Eq. (1) was rewritten as

$$\delta^2 = (\delta')^2 + m\alpha\beta \quad (9)$$

where m was an adjustable parameter. Minimizing the objective function $\sum\{(\delta')^2 - (\lambda^2 + \tau^2/2)\}^2$ for 30 compounds in the database of Lazzaroni et al. where δ' was computed from Eq. (9) yielded an optimal value of $m \approx 2.2$. The factor of 4 in the denominator of Eq. (4) was determined by minimizing deviations in vapor-liquid equilibrium (VLE) data for the 10 binary systems of the first five compounds in Table 1, and three VLE systems involving water with special emphasis on ethanol+water. These particular mixtures were chosen to illustrate the range of possibilities from strong solvation to hydrophobicity. Coincidentally, this manner of separation retains the consistent interpretation of cohesive energy density as a primary consideration (e.g. $\delta_W' > \delta_B' > \delta_O'$, where W means water, B means benzene, and O means iso-octane). This consistency is not immediately apparent in MOSCED's λ parameters.

Other simple alternatives include the Hansen solubility parameters[4,5] and the original Scatchard-Hildebrand model.[5,6] The problem with the Scatchard-Hildebrand model is that it overestimates the non-ideality of the solution. For example, matching the experimental data for methanol+benzene requires a negative value of k_{12}= -0.035 when the Scatchard-Hildebrand model is applied. Students then conclude that methanol and benzene must "like" each other, because that is what k_{12} should mean. But in this case, the negative k_{12} is cancelling the overestimation of the non-ideality from the "unseparated" cohesive energy density. A similar problem occurs with water and nearly any other compound. Even ethanol+water is predicted to be immiscible with the Scatchard-Hildebrand model. This kind of "two steps forward and one step backward" makes the presentation unnecessarily confusing. The Hansen solubility parameters, on the other hand, have the advantage of being simpler than SSCED in some sense, because there is a single hydrogen bonding parameter instead of two. Nevertheless, Hansen's method cannot account for activity coefficients less than one. This undermines the scope of conceptual reasoning that should form the long term basis for students' thermodynamic insight. In deference to Hansen's method, however, the separate contributions to the solubility parameter are constrained to sum to the original Scatchard-Hildebrand value. This adaptation from Hansen's method is helpful in clarifying that SSCED provides separation, but no elimination.

An advantage of the MOSCED model is its direct accounting of the specific molecular interactions involved in hydrogen bonding. This accounting is based on spectroscopic measurements that are independent of the desired activity coefficients.[7] Kamlet-Taft parameters are dimensionless measures of acidity and basicity, but the MOSCED model recasts their values to provide dimensional consistency with solubility parameters. This may open the door to more creative ways for students to mesh analytical techniques with engineering applications, as in catalysis for example. Interactions of zeolite acid sites with molecular base sites may seem less mysterious when the existence of molecular base sites has been acknowledged at the outset. This improved chemical insight can be pervasive throughout the curriculum.

Eq. (1) was deliberately expressed in terms of k_{12} instead of simply substituting Eq. (4) directly, as in MOSCED. Note that Eq. (4) is described as a "guideline." This means that it is a starting point, but it leaves open the possibility of refining the value. The value of experimental data is hinted at in the presentation and several assignments at the sophomore level lead students through the process of finding relevant data and inferring refined values of k_{12}. A model like UNIFAC can predict activity coefficients but sheds no light on the underlying chemical interactions that lead to the behavior. Furthermore, the UNIFAC model makes it difficult to refine predictions in light of experimental data for specific systems of interest.

Another alternative would be Wertheim's theory.[8] Wertheim's theory forms the basis of hydrogen bonding equations of state like the PCSAFT[9] and ESD[10] models. It is also based on rigorous statistical mechanics instead of phenomenological arguments. Therefore a simplified integration of Wertheim's theory with Scatchard-Hildebrand theory would have an advantage as a natural segue to the more sophisticated theories. In fact, this was considered to be the preferred alternative initially. Unfortunately, hydrogen bonding compounds tend to operate at higher packing fractions, and this undermines a simple extension of the constant packing fraction assertion. Some compromise may be feasible long term, but the current form of the SSCED model is satisfactory for present purposes.

Finally, it may be possible to relate the dispersion, acidity, and basicity parameters to ab initio characterizations, as in the COSMO-RS model.[11] This would reinforce the value of quantum mechanical computations throughout the curriculum, but freshmen (and sophomores) would be unlikely to appreciate this level of sophistication. It would make more sense to recast the ab initio results as reinforcing the SSCED concepts after the fact, in the Junior or Senior year.

There is one substantial disadvantage of the SSCED model that motivates the coverage of more advanced models like UNIQUAC and SAFT. Since it has only one binary interaction parameter, the magnitude of deviations from non-ideality can be adjusted, but not the skewness. The skewness of the Gibbs energy is controlled by the volume ratio in the SSCED model. However, this limitation pertains to quantitative modeling, not to the conceptual and educational device intended here.

4. Assessment

Assessments of student learning have not been performed for the freshmen yet, but a very similar presentation pertains to the sophomores and this has been assessed in class through the ConcepTest methodology and through traditional examination questions. These are less personal than looking students in the eye to gage their understanding, but they do provide objectivity.

ConcepTests pose simple questions to the class and allow them to post their answers anonymously for quick compilation.[2] Electronic devices typically facilitate this approach, but it can be conducted with colored flash cards. In the strictest sense, ConcepTests should focus entirely on conceptual questions, but a small adaptation permits engagement in active learning for computational exercises as well. I refer to these as CompuTests. A few examples are given below. The (%) quantities refer to the % of students who answered correctly.

ConcepTest 1.

Referring to the figure below, cases A, B, and C correspond to characterizations of the attractive energy between two molecules as described by the square well potential. This attractive energy is given by $\varepsilon_{12} = (\varepsilon_1 * \varepsilon_2)^{1/2} (1 - k_{12})$ where ε_{12} gives the depth of the square well. Note that the bottom of the well is $-\varepsilon_{12}$. Provide the response (A or B) corresponding to each situation.

a. Which (A or B) corresponds to $k_{ij} > 0$? (72%)
b. Which corresponds to components "liking" each other? (100%)
c. Which corresponds to a higher "escaping tendency" for component 1? (94%)
d. Which will give the highest bubble point pressure? (50%)

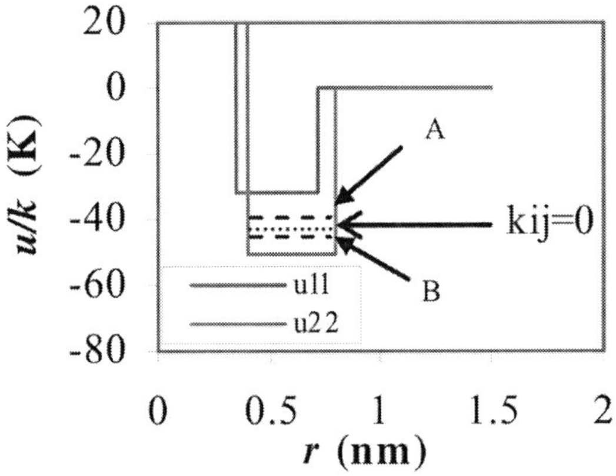

ConcepTest 2. (94%)

Which of the following properly depicts a maximum boiling azeotrope?

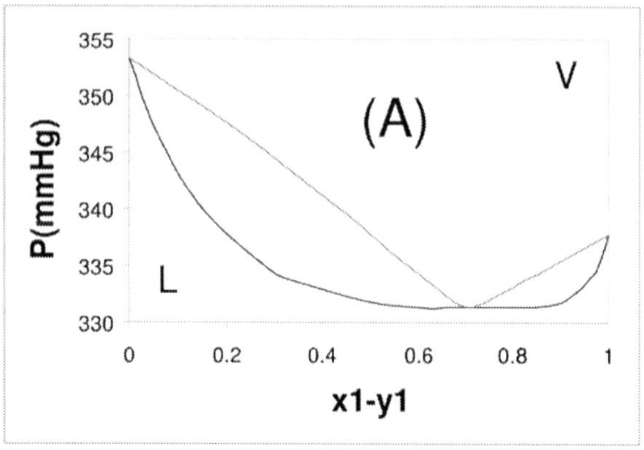

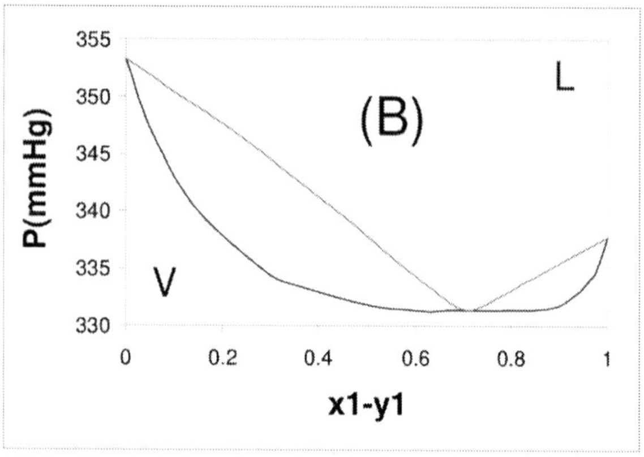

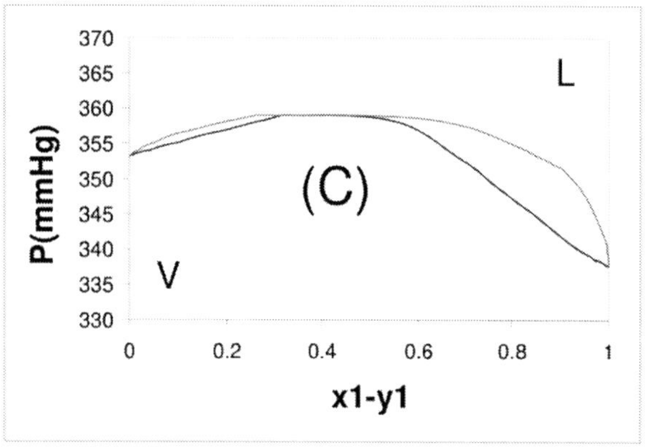

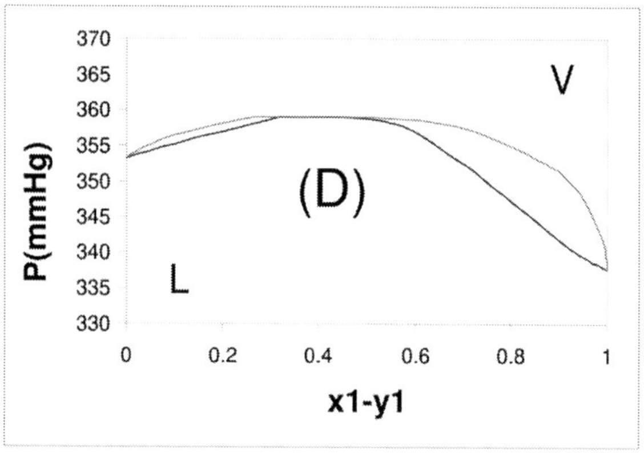

"CompuTest" 1. (43%)

Arrange the following mixtures from most compatible to least compatible according the SCED solubility parameter criterion (k_{12}=0). (1) Pentane+hexane, (2) decane+decalin, (3) 1-hexene+ dodecanol, (4) pyridine+methanol, (5) diethyl ether+n-heptane

A. 12345 B. 12534 C. 54123 D. 21543

"CompuTest" 2. (76%)

An azeotrope exists for n-butane(1)+ethyleneOxide(2) at 1.013 bars at -6.5°C and 78wt% butane. Estimate the activity coefficient of EtO (γ_2) at the azeotropic composition and temperature from the Scatchard-Hildebrand model assuming k_{ij} = 0.

Compound	Tc(K)	Pc(MPa)	w	CpIg/R	MW	δ(cal/cc)$^{1/2}$	ρ298
n-BUTANE	425.2	3.80	0.193	11.89	58	6.60	0.60
EtOxide	469.0	7.10	0.200	5.80	44	10.62	0.89

(a) 0.04 (b) 1.06 (c) 1.98 (d) 2.89

Examination Question 1. (89%)

Based on the Scatchard-Hildebrand solubility parameters (k_{12}=0), arrange the following mixtures from most ideal to most non-ideal: (a) 2-pentanone+1-pentene, (b) 2-pentanone+ naphthalene, (c) ethanol +naphthalene, (d) n-hexane+ ethanol.

Examination Question 2. (61%)

A common problem with recycling polyester is the impurities from bottle caps and labels. The bottle caps typically weigh 0.05 g and the bottles are 2g. The caps are polypropylene (PP) with molecular weight of 60,000 g/mol. The bottles are polyethyleneterephthalate (PET), with molecular weight of 10,000g/mol. The solubility parameter and density of PP can be estimated from those of 224TriMethylPentane. The solubility parameter and density of PET can be approximated with those of p-xylene. Estimate the infinite dilution activity coefficient for PP in PET at 100°C assuming $k_{ij}=0$.

5. Conclusions

Two issues pervade teaching in the chemical engineering curriculum: time constraints and knowledge retention. If you teach too much in too little time, little is retained. If you teach too little, students cannot "connect the dots" from one isolated fact to another. The key is to articulate broadly applicable concepts, like the SCED model presented here. This mindset leverages familiar chemical concepts like acidity and basicity while dovetailing nicely with the physical interactions of the van der Waals model covered in physics coursework. Leveraging the concepts presented in other coursework has the two-fold advantage of saving time and rewarding students for retaining what they learn from course to course.

The assessments show that this perspective is accessible to sophomores at least. Assessments were not conducted for freshmen at this early stage of adapting the presentation for them, but we expect similar results to those for sophomores if sufficient effort is devoted. Students are able to quickly recognize solution non-ideality and the impact this may have on solubility and volatility. Students whose careers take them away from process design may not remember how to compute an activity coefficient five years after graduating, but they can remember that acids and bases interact strongly and that even organic chemicals can be formulated to take advantage of those interactions.

Reference:

1. M. J. Lazzaroni, D. Bush, C. A. Eckert, T. C. Frank, S. Gupta, J. D., and J. D. Olson, Ind. Eng. Chem. Res. **44**, 4075 (2005).
2. J. L. Falconer, Chem. Eng. Ed. **38**, 64 (2004).
3. J. R. Elliott, Chem. Eng. Ed. **27**, 44 (1993).
4. C. M. Hansen, *Hansen Solubility Parameters: A User's Handbook.* (CRC Press, Boca Raton, Fla, 2000).
5. G. Scatchard, Chem. Rev., 321 (1931).
6. J. H. Hildebrand, J. M. Prausnitz, and R. L. Scott, *Regular and Related Solutions.* (Van Nostrand-Reinhold, New York, 1970).
7. M. J. Kamlet, J. M. Abboud, M. H. Abraham, and R. W. Taft, J. Org. Chem. **48**, 2877 (1983).
8. M. S. Wertheim, J. Stat. Phys. **35**, 19 (1984).
9. J. Gross and G. Sadowski, Ind. Eng. Chem. Res. **40**, 1244 (2001).
10. J. R. Elliott, S. J. Suresh, and M. D. Donohue, Ind. Eng. Chem. Res. **29**, 1476 (1990).
11. A. Klamt, J. Phys. Chem. **99**, 2224 (1995).

AUTHOR INDEX